OUT OF THE SCIENTIST'S GARDEN

OUT OF THE SCIENTIST'S GARDEN

A STORY OF WATER AND FOOD

RICHARD STIRZAKER

CSIRO
Land and Water Australia Senior Research Fellow
CRC for Irrigation Futures

CSIRO
PUBLISHING

National Library of Australia Cataloguing-in-Publication entry
Stirzaker, Richard John.

Out of the scientist's garden : a story of water and food / Richard Stirzaker.

9780643096585 (pbk)

Bibliography.

Crops and water.
Gardens – Irrigation.
Agriculture.

630

Published by
CSIRO PUBLISHING
150 Oxford Street (PO Box 1139)
Collingwood VIC 3066
Australia

Telephone: +61 3 9662 7666
Local call: 1300 788 000 (Australia only)
Fax: +61 3 9662 7555
Email: publishing.sales@csiro.au
Web site: www.publish.csiro.au

Front cover photo by iStockphoto

Back cover photo by Brad Collis

Set in 10.5/16 Adobe Myriad MM and Optima
Edited by Janet Walker
Cover and text design by James Kelly
Typeset by Desktop Concepts Pty Ltd, Melbourne
Printed in China by 1010 Printing International Ltd

The paper this book is printed on is certified by the Forest Stewardship Council (FSC) © 1996 FSC A.C. The FSC promotes environmentally responsible, socially beneficial and economically viable management of the world's forests.

CSIRO PUBLISHING publishes and distributes scientific, technical and health science books, magazines and journals from Australia to a worldwide audience and conducts these activities autonomously from the research activities of the Commonwealth Scientific and Industrial Research Organisation (CSIRO).

Contents

Prologue: Entering the scientist's garden

One of the first things I learned in science class at primary school was the difference between renewable and non-renewable resources. Oil was a non-renewable resource. One day we would suck the last barrel out of the earth and there would be no more. The oil price skyrocketed through the 1970s and has been doing so again, along with predictions about how many years' supply we have left.

Water was different. It was continuously renewed in an endless cycle of evaporation and rainfall. A picture in my school textbook showed wavy arrows of moisture rising up from the sea, feeding huge clouds that rained over the land. Forests grew from the wet earth, returning water to the sky. Rivers carried the remainder back to the ocean. Burn oil and it was gone forever, but water keeps evaporating, condensing and evaporating again. Ultimately, it cannot be consumed.

Oil will run out someday, but so far no petrol station has stopped me filling my car. I am, however, forbidden to turn on my garden tap and water the tomatoes in my back garden – at least on every second day. The great renewable resource is not being renewed fast enough where I live.

I was born in the city of Cape Town, at the southern tip of Africa. Through my growing years, I answered with ease the question most asked of children by adults: 'What do you want to do when you grow up?' I wanted to be a farmer.

My desire to farm started with a primary school experiment. We each had to bring some seeds to school, and from the day mine germinated I was hooked. I started a vegetable garden and have grown one ever since, first in Cape Town where I grew up, then in Sydney through my university years and now in my home town of Canberra.

I never farmed, but my job as a research scientist revolves around turning water into food and how the wider environment is affected by the way we farm the land. I also spend part of each year running experiments and teaching postgraduate students in Africa. Over a couple of decades I have witnessed all means of applying water to crops – from the latest drip irrigation technologies to women and children carrying water in rickety carts drawn by donkeys.

Through all this I have maintained my own fruit and vegetable garden, and it has become a laboratory and a sanctuary; a place to think and learn about how we use water to grow food. When I got the chance to write this book through a fellowship from the research corporation Land and Water Australia, I knew immediately what I wanted to say.

My garden is the place where I mull over the things that have troubled me during the working day. My concerns coalesce around four issues. First, it is surprisingly hard to use water well. We have become used to having lots of it and we are not well-placed to face a world of sudden scarcity.

Second, the way food is produced has changed dramatically in the last 50 years, with most of our food grown by very few people on large, specialised farms. Unlike the garden, the modern food chain is long and complex, and supplying the supermarket shelf uses far more fossil fuel energy than we derive from eating the food we buy.

Third, my time spent in Africa has exposed me to the precarious life of the small-scale farmer. We do not have enough ideas to lift the productivity of the resource-poor farmers, and the neediest of all are hard to reach with the ideas we do have.

Finally, I am troubled by my own profession. We seem to be drifting into a culture of short-termism and managerialism, just at the time we need to nurture the creativity and boldness required to face up to a bewildering array of global problems.

My troubles are only eclipsed by the wonder of a garden. At home we call the month of August the start of the hungry gap. We have finished the fruit preserved from summer, emptied the store of onions and potatoes and eaten through the greens planted the previous autumn. The garden beds are cold and bare. But then comes the renewal of spring. The fruit trees burst into blossom and the crops emerge from the warming soil. Soon the first produce is ready for harvest – food and flavours of which I never grow tired.

This book is written for anyone who wants to understand water a little better – for those growing vegetables in a garden, food in a subsistence plot or crops on vast irrigated plains. But it is also for anyone who has never grown anything before and who has wondered at the passion of the gardener or farmer. It is also for anyone who shares the troubles I have outlined above, of a world with shrinking resources feeding a growing population.

The book is divided into three sections, each containing seven essays. Part 1 'The view from our garden' contains stories about why things work the way they do, through the lens of my backyard fruit and vegetable garden. Part 2 'A journey through soil' explores irrigation, soil, land, rivers and aquifers and how to navigate the fine line between productivity and degradation. Part 3 'Feeding ourselves' takes you on a brief history of agriculture, discusses how the world feeds itself and explores some of the conundrums of modern agriculture.

Each section deals with soil, water and food from different perspectives and scales, but you will quickly see that they are all part of the same tapestry. Each essay is a story in itself and the book can be read in any order, but I think you will find that each builds on the one before, creating a bigger story – the story of turning water into food.

Acknowledgements

I wrote this book whilst on a fellowship provided by Land and Water Australia. The intent of the fellowship was to sponsor a reflective piece of work by 'freeing up' the time of a mid-career scientist from the constraints of their everyday chores. I enjoyed every minute of this fellowship, and am incredibly grateful to Land and Water Australia and to CSIRO, my employer, who supported it all the way through.

For most scientists, writing is an important part of their job. But we write for academic journals – a terse and precise form of writing, painstakingly argued and meticulously referenced. This book could not be more different. I had a free hand to explore ideas and link them up in ways no journal editor would ever allow. The content of the book is drawn from my own experience, and from many colleagues who have shared their insights in their own writings and in countless conversations. You are not referenced here, but I trust you will recognise your contributions as you read through this book.

Like all scientists, I am indebted to the agencies that have funded my work, in particular Land and Water Australia, the Rural Industries Research and Development Corporation, the Water Research Commission of South Africa and the CRC for Irrigation Futures.

Ted Lefroy and John Passioura acted as my science reviewers and David Salt helped pull the final version together. My gratitude to them extends way beyond those tasks, for I have enjoyed their friendship and wisdom over many years.

The obligate gardener is not always easy to live with. They do not want to go on holidays in summer when the crops are in full swing. They get anxious when they miss sowing dates, are downcast by outbreaks of pests and disease and edgy at the prospects of late frosts or summer hail storms. Yet my parents encouraged me all the way as I slowly took over their garden. My own family is incredibly tolerant with the one who won't come inside until the sun goes down. I am especially thankful to my wife, Mary, for her faithful support and wise counsel through many of the stories recorded in this book.

PART 1
THE VIEW FROM OUR GARDEN

This is not a book about gardening. Yet the story of feeding the world begins in a garden. The first farmers were hunter–gatherers who started by growing food around their temporary dwellings. These days few of us grow food for a living, but most of us have grown food in a garden or know somebody who does.

Talk about feeding the world invokes numbers that are hard to conceptualise – hundreds of millions of people, thousands of tonnes of food and gigalitres of water. But the garden is a scale we can comprehend, and throughout this book the garden is the lens through which we will explore how the world feeds itself.

This first collection of seven essays is drawn from experiences in my own garden and charts my journey from growing a few salads to growing vegetables on a commercial scale. It starts with the big picture on water, how much there is and how much we need. Then we meander through the vegetable garden and the orchard, before moving on to soil fertility, pests and diseases, biodiversity, and what makes us crave certain foods.

I spent my growing years planting every sunny spot in the garden with vegetables. *The Complete Book of Self-Sufficiency* was my constant companion. Then I studied agriculture, started a career in research, and filled every garden I lived in with edible plants.

Slowly more and more fruit crops were integrated into the garden. Growing fruit is a bit harder than growing vegetables. You need longer term plans and there is more that can go wrong. Sometimes you get fantastic rewards for doing almost nothing.

A garden becomes a small ecosystem that attracts other life – both friend and foe. Soil is full of micro-organisms which decompose old vegetation and recycle nutrients for the new crops. Silvereyes and Willie Wagtails scout for pests. Meanwhile fungi rot the ripe fruit in wet weather, caterpillars chomp through leaves, and mites and aphids suck their sap.

My first job involved growing vegetables on a commercial scale and I was faced with all kinds of problems my *Complete Book of Self-Sufficiency* had not prepared me for. Real farming was not just a very big garden, and getting uniform blemish-free produce to market posed a new set of challenges. I developed a huge respect for farmers.

Chapter 1

The colour of water

I live in the catchment of the Murray–Darling River system, the food bowl of Australia. This is a country of droughts and flooding rains, which is why there are enough dams to store twice the volume of water that the river system carries in an average year. It is said that if all the storages in Australia were full, each one of us would have enough water for three Olympic-size swimming pools.

Somehow we are getting through our Olympic-size swimming pools, at least in the southern half of the country. Demand for water by irrigators has steadily increased, as has the requirement from the cities, while the long, slow drought drags on. Almost everyone is on water restrictions. For the first time ever, irrigators are bulldozing out their fruit trees. Horticultural businesses in the cities have gone broke. It will rain again, and for a time the dams will be filled. But water shortages will not go away for long. So what have we learnt that will prepare us for a world of scarcity?

There is plenty of water on earth, but 97% of it is in the oceans. Each litre of sea water contains around seven teaspoons of salt, most of which is sodium chloride or common table salt. The concentration of salt in the sea is about four times greater than the concentration of salt in our blood. If we drink sea water, the salt concentration of our body and then our blood goes up. In an attempt to dilute the extra salt in the blood, water is extracted from our tissues, causing us to dehydrate. So it's true; the more sea water you drink, the thirstier you get.

Plants have a similar problem. Irrigate with sea water and plants will quickly wilt, as they do in dry soil. Less than 3% of water on earth is fresh enough for watering plants, and two-thirds of this is locked up in frozen icecaps. Most of the rest is present in aquifers deep below ground. Just one-tenth of 1% of the fresh water supports biological life, which includes the water in lakes, rivers, dams, soil and vegetation. Accessible fresh water is a limited resource.

There is a water tank on the side of our house which is about 1.8 m high and 1.8 m in diameter and holds 4500 litres of water. If I lived in an African village, I might get my fresh water from a communal tap in the street. Say it was a short three minute round trip to fill, carry and empty a nine litre watering can into the tank. I would need to make 500 round trips to fill a single tank, and with no meal or sleep breaks it would take 25 hours. My whole day would be used up carting water to fill one tank.

One 4500 litre tank is about the amount of water I, as an average westerner, will consume today. How could one person drain a full water tank in just one day, a tank that took longer than a day to fill?

The average Australian used around 280 litres of water in the home every day in 2005. It sounds a lot of water, but it is 50 litres less per person than was used just four years previously. Of this 280 litres, 123 goes to watering the garden, 56 litres for washing ourselves and 36 litres for washing our clothes. We use 23 litres in the kitchen and flush 42 litres down the toilet. These are, of course, averages. Some people do not have gardens at all, or don't water their garden.

After removing my 280 litre daily domestic requirement from the tank, there are still 4220 litres left. Where does this go? From here on the numbers get a little rubbery. Water is used in everything we need for daily life, from mining to manufacturing to power generation. Anything made of metal, paper or plastic has used water in the manufacturing process. Based on figures for the economy as a whole, it is a reasonable guess that at least 220 litres of water was used to provide me with these products and services.

That leaves 4000 litres in the tank, and this water is used up producing the food I eat every single day. This is the focus of this book. Most of the water we get through each day was used in producing the food we eat.

A patch of crop occupying 10 paces by 10 paces can easily use 500 litres of water on a warm summer's day. That means a half a ton of water, so it is worth knowing what it's used for. Plants themselves contain around 90% water, but only about 2% of the water that enters the roots is retained by the plant in the cells of new shoots and leaves. The rest just goes through the leaves and back to the atmosphere as transpiration.

The transpiration stream is the movement of water from the soil, through the roots, to the stem and through the leaves back to the air. Water evaporating from leaves helps to keep them cool. Transpiration also draws nutrients, dissolved in the water, into the roots. Yet the 500 litres used by the crop is more than is required for cooling and nutrient supply. Most of the water is lost as a consequence of the business of photosynthesis.

Photosynthesis takes place in leaves when carbon dioxide (CO_2) is turned into sugar under the power of sunlight. Below the waxy surface of the leaf, enzymes are busy combining water with six CO_2 molecules to make a sugar molecule inside water-filled cells. The CO_2 has to get from the atmosphere to the inside of the leaf, and this occurs through small pores in the surface of the leaf which the plant can open and shut.

There are air spaces between the watery cells inside the leaf. This air is near 100% humidity, but is constantly stripped of its CO_2 to feed photosynthesis. The pores on the leaf surface are opened to let in more CO_2-rich but dry air from the atmosphere. At the same time the humid air inside the leaf escapes through the same pores. New water must be pulled up from the roots to replenish that which was lost from the leaf. If there is insufficient water in the soil, the plant will close its leaf pores to reduce its water loss, otherwise it will wilt. At the same time it shuts out its supply of CO_2 from the air. A plant without water is starved of food.

All plants produce sugar from sunlight using CO_2 as the raw material and producing oxygen as a by-product. Sugar provides the building block for carbohydrate, fat and protein. Animals rely totally on plants for their supply of food, releasing food energy through the process of respiration. Respiration is essentially the reverse of photosynthesis. The sugar from carbohydrate and fat is burnt in the presence of oxygen to fuel our bodies, with CO_2 as the by-product. CO_2 is cycled round and round, from air through photosynthesis to plant to animal and through respiration back to CO_2. Burning the products of ancient photosynthesis like coal and oil, is of course, a different matter.

An adult must eat around 2500 calories of food energy to power the body for a day. As a rough guide, it takes at least one litre of water to produce one calorie of food. The actual amount of water needed to produce a calorie depends on the kind of food we eat. For example, wheat gives us more calories per litre of water than rice. This is because wheat is grown during cool winters, while rice is grown in warmer climates where there is more transpiration.

Animal products require more water per food calorie than wheat or rice. It takes around 10 calories worth of grass to make one calorie of meat or milk. This is because a substantial proportion of the food energy is needed to keep animals alive before they can convert what they eat into meat, milk or eggs.

Someone on starvation rations would eat food that used the equivalent of 1000 litres per day to grow. A vegetarian could eat their way through around 2500 litres, but the typical red-meat Western diet can get through twice that much. It is not hard to use the remaining 4000 litres in the tank.

'Blue water' is the water that runs off the soil surface and reaches a river, or percolates through the soil to an aquifer. Blue water can be collected, stored and pumped to our homes, to factories and to farms.

There is a twist to these figures. The water used to produce our food does not all have to come from the water tank. Much of our food is produced from crops grown with rainfall alone. This is where the colour of water comes in.

Rain that falls on the earth and replenishes the soil's store of water we call 'green water'. It can only be used by plants actually growing in the place where the rain fell. If water runs off the soil surface and reaches a river, or percolates through the soil to an aquifer, we call it 'blue water'. Blue water can be collected, stored and moved to any sector of the economy, or left for the environment. For example, we can use blue water to provide piped water for sanitation, or use this water for irrigation.

If a cow ate grass that grew under natural rainfall, then all the water consumed is called green water. Green water is water held in the soil and it cannot be used for any other purpose, except by the plants actually growing there. If we replace animal protein produced under rain-fed conditions with soybeans grown under irrigation, we use blue water instead of green. Blue water could have been used for some other purpose – in homes, manufacturing or left in the rivers. So although it is good to eat less animal products, it is not the whole story.

As the population increases and as diets change, the demand for water goes up. Over the last 40 years, the amount of calories in our diet provided from animal products has doubled, requiring more water. This is a global average, and reflects the fact that many countries have large segments of the population becoming more affluent and upgrading their diets.

Of all the fresh water humans remove from rivers, dams and aquifers – the blue water – 70% is used to irrigate plants. The remaining water is used in homes and in the industrial sector. The trend is towards less water being available to grow food. This is because domestic and industrial users are willing to pay far more than farmers can, and will take a bigger share.

It is obvious that water is becoming a scarce resource, but exactly how scarce is surprisingly difficult to answer. Some areas are physically short of water right now; there is not enough fresh water to satisfy the demands of the population. A sweep of countries from Morocco to Egypt in North Africa, through the Middle East and into central and west Asia, fall into this category. In the next decade or so, eastern and southern Africa will join them. In other areas, like central Africa and South America, water is economically scarce. There is water available, but not the infrastructure needed to bring the water to the people who need it.

Right now, one in six of the world's population drink from the same water source they use to wash their clothes and water their animals. Similarly, 40% do not have access to improved sanitation, which prevents human effluent contaminating drinking water. Poor communities bear a heavy burden from the plethora of water-borne diseases and parasites under these conditions. Providing running water and improved sanitation is technically straightforward and relatively inexpensive in a global context, so it remains a conundrum why progress on this front has been so slow. One thing is for sure; some of the available blue water will have to be redirected to this, and away from irrigation.

This brings in the remaining two colours of water. Grey water is the water released from our baths, showers and laundries. Black water comes from the toilet. These two colours also highlight some of the difficulty in working out how much water is available. The same litre of water may be used for hydro power generation, domestic water and finally irrigation through a water recycling scheme. Or black water may contaminate blue water, when a previously clean water source becomes polluted by sewerage.

Finally, there is the curiously named 'virtual water'. When a country imports food it effectively imports the water that grew the food. So a country may appear to be water scarce, but in fact has all the water it needs because it does not need to grow as much food. Countries like Australia that export a lot of food are therefore also exporting water.

As a young backyard vegetable grower, I watered my plants if it had not rained for a while, or if the ground looked particularly dry or the plants wilted a bit. If I harvested a couple of kilos of tomatoes from a square metre of garden I thought I was doing pretty well. It was only when I started growing vegetables on a commercial scale as part of my postgraduate studies that I realised that my best garden yields were half that considered acceptable by a farmer, largely because of my haphazard watering strategy.

I did produce 10 kg of tomatoes per square metre in a commercial crop, but these were fertilised heavily and drip irrigated just about every day. After this experience I approached my suburban garden in a totally different way. Proper irrigation required a fair amount of water, and it seemed a shame to use only chlorinated and filtered water from a tap, when the water falling on the roof and going down the plug hole was wasted. In those days it was illegal to install a rainwater tank in Canberra, and unthinkable to use grey water. Then, five years later, the same water authority offered to pay a substantial rebate if we did install rainwater tanks. Water was running out.

If I was a very good tomato grower, I could produce this basket full of tomatoes on one square metre of land, using about 500 litres of water in the process.

For decades there had been one pipe carrying an endless supply of clean water into the house and one pipe carrying the dirty water away. The engineers who had designed and built the system did not want anyone tampering with it, not even with the storm water. But the great engineering solution that has brought so much benefit to humankind required a rethink. The blue water supply was no longer looking endless and many were looking to the grey.

City water restrictions have been with us for a number of years now and it is the gardeners who are hardest hit. There are no water restrictions for inside the house, but the garden can only be watered between certain hours on certain days and, for the most part, must be carried out by hand or through drip irrigation. The gardeners of my city, Canberra, proved to be very cooperative. In the three years prior to restrictions, the yearly average household water consumption in the city was 320 000 litres. In the first three years after restrictions it had dropped by over 20%, mostly because people stopped watering their gardens.

In the three years before water restrictions, our household water use was about one-third higher than the city average. This is not as bad as it sounds. We have a large productive garden covering every square metre of the property. The city average includes all those who live in apartments, and have no garden at all. We also have five occupants in the house, including two teenagers, whereas the average occupancy is around two-and-a-half.

So when the water restrictions hit, we had to make a lot of changes. We installed three water tanks, set up a grey water system, redirected all run-off from hard surfaces into soak-away trenches, and installed all kinds of soil water monitoring equipment in the garden beds. We were determined to keep a production garden going, despite the restrictions. Over the last five years we have used 35% less water from the city supply and kept our garden just as productive through the drought.

We have not thrown out the dishwasher or put a timer in the shower, but we have had a lot of fun saving water in the garden. The drought is still hanging on; average rainfall over the past five years has been 110 mm less per year than the long-term 630 mm we were used to. Maybe we will creep back towards average rainfall. In the meantime, saving water has become a passion. If the average domestic water consumption is 280 litres per person per day, then our family gets by on half that while producing the bulk of our fruit and vegetables in the middle of town. We have learned to make the most of the green water that falls on the garden, the blue water collected from the roof and the grey water from the bathrooms and laundry.

In short, water restrictions have been a blessing in disguise. In my job I was always telling other people how to use less water, but now I have to do it too. The garden has become a testing ground, a laboratory of ideas, and has even spawned new inventions. It has been a journey of learning that has helped me in my work, among some of the biggest irrigators in the country all the way to small-scale farmers in Africa.

Chapter 2

A year in the vegetable garden

When I started growing vegetables as a boy, I pored over the 'sowing dates' page in my well-worn vegetable growers' handbook. The months of the year ran along the top of the page and all the different vegetable crops ran down the left-hand side of the page. The correct months for sowing or transplanting were marked with a big dot. If you missed the sowing date then you had to wait a whole year. 'What should I be planting now?' is still the most common question I get asked.

The problem with the sowing date chart is that you tend to plant rather sporadically and then have short-lived gluts a few months later. We buy a punnet of cucumber seedlings in late spring, and by mid-summer we are picking three cucumbers a day. Picking can be one of the most time-consuming chores in the garden and so gluts are a problem. We pick a bag of beans every second day and shove them in the bottom tray of the fridge. It is demoralising to find a number of rather unappetising bags still there a month later after eating beans with every meal.

During the peak harvest times of late summer and autumn, it is not unusual for freshly picked vegetables to spend a fortnight in the fridge and then go straight into the scraps bucket and out to the chickens. There is just too much to eat. During late winter and spring there are slim pickings, or at least not much variety to choose from. This is the challenge of growing vegetables in a place with a fairly long winter and a short, hot summer.

Each vegetable crop has an optimal sowing date, but for many crops there is quite a bit of leeway either side. We call this the sowing window. If you want to spread the harvest throughout the whole year, then you need to be able to push the sowing windows to their extremes. It is better to think about what we will have to harvest during each week of the year, and then work backwards to what we need to plant.

The secret to exploiting the widest possible sowing window is to know what limits survival and then productivity. Once we know this, we can work out the earliest and latest date we can plant tomatoes, or the best time to sow onions.

Vegetables fall into two groups: warm season crops that are killed by frost and cool season crops that can tolerate frosts and benefit from a period of cold weather. The sowing window for warm season crops starts when the soil is warm enough for the seed to germinate and closes as the cold late autumn weather takes hold.

Plants that tolerate cool weather also fall into two groups. The cabbage family grows best when it starts in the warmth of late summer and then matures into the winter. The onion family likes to start in cooler weather and mature into the warm weather. Basically cabbages mature as the day length gets shorter and onions as the days get longer. Get this wrong, and the crop runs to seed before there is anything to harvest.

Frost is the great show-stopper. Most of my gardens have been away from the moderating influence of the coast, so the last frost of spring and the first frost of autumn define the growing season for all the warm season crops. Of course it's not possible to know which frost will be the last, so it is always a gamble. It is easy to get fooled by an unusually warm week in early spring, and most keen gardeners have a date in their mind before which they are not willing to take the risk.

It is good to have these dates in mind, but the young seed or seedling responds not to a particular day of the year but the temperature of the soil. The seed imbibes water as soon as we place it in the earth. Immediately, enzymes get to work on the stored energy in the seed and the primordial shoot and root start to develop. The speed of each enzyme reaction depends on the temperature and this is crop dependent. If you plant a pea seed and a tomato seed in soil at 10°C, the pea will germinate within a couple of weeks but the tomato not at all, regardless of what day of the year it is.

When the soil warms to 15°C, beans, beetroot, carrots and all the cabbage family will germinate easily. It is best to wait until the soil temperature rises to 20°C before planting tomatoes, capsicums, eggplants, cucumbers, melons and corn. Leeks and onions, which are planted in the autumn or winter, will germinate in soils at 10°C, and broad beans and peas in soil as cold as 5°C.

In the home garden we need to jump-start the season, and we do this with seedlings. I start my tomato seedlings on a heat mat in my glasshouse in late winter while it is still cold and frosty outside. Then I transplant them into individual pots, and by the time I'm ready to plant outside after the last frost, the plants are growing strongly with the first flowers already open.

It is easy to fill the garden in summer, have a food glut in autumn and run out of vegetables in winter. Mid to late summer is an important time to get the cool season garden started before the cold weather sets in.

I grow capsicums, cucumbers and zucchinis in large pots before planting outside so I have fairly big plants to take advantage of the warmth of spring when it finally sets in. In the garden there are some raised beds with brick walls and the soil warms up in these before any other beds. By getting the plant as big as possible in the glasshouse, planting out as soon as I dare in spring into the warmest parts of the garden, and being prepared to cover the plants at night if a frost threatens, I can start the harvest two months earlier than if I waited for the soil to warm to the right temperature.

After the very early crops are in the ground, I can think of planting the main warm season crops as I clear out beds that grew winter vegetables or cover crops. Some of the summer crops, like capsicums and pumpkins, will keep producing right through until the first frost of autumn. Other crops, like zucchinis, cucumbers and some tomato varieties, tend to get overtaken by pests and disease with age. It is best to have sequential sowings of these, and to pull out the earlier plantings as they lose their vigour. This time we are not gambling on the last frost, but the first frost of the coming winter.

Most gardeners feel the sap rising in spring and are itching to get their vegetable patch moving, but the late summer/early autumn planting period is just as important, and much easier to miss. The garden is so full of vegetables in late summer that there may be no space to plant. But this is the very time to be preparing for the long winter ahead. There are a whole lot of vegetables that need to do most of their growing in the late summer warmth, and then mature into the colder months. The most important of these are the brassica family – cabbage, broccoli, cauliflower, Brussels sprouts and kale.

Cabbage and broccoli can be grown through spring and summer, but it is not worth the bother. First, there is so much other stuff to eat at this time of year. Second, you'll be in an endless battle with the cabbage white butterfly and some other persistent caterpillars. Third, you are going to be on a steady diet of brassica vegetables from winter to spring, because there is little else to eat.

In our climate it is essential to have the first brassica seedlings into the garden before the end of summer and any plantings after mid-autumn are dubious. We don't want to plant too early for the reasons above – the battle with caterpillars, which is just about over when the frosts hit – and because there is so much other stuff to eat. But if we plant too late, and the plants are half grown by the time the cold weather sets in, then the harvest is pitiful. Brassicas, like all vegetables, must grow fast. We must get them to the stage where adjacent plants are barging each other out of the row while there is still warmth in the autumn sun. The plant must grow a big frame if we want to harvest big heads of cabbage, broccoli and cauliflower.

There is a big difference between sequential planting in spring and sequential planting in autumn. Since the weather is warming up in spring, the later planted vegetables tend to catch up with the earlier planted ones. Tomato crops planted six weeks apart in spring may only fruit two weeks apart. We see the opposite in autumn because the weather is getting colder. If you plant a month apart the harvest dates could be two months apart, because the earlier planted crop gets its start in the warmer weather. In fact the later planted crop may not get to a decent size before the cold cuts in, and we may get no harvest at all.

Although it is useful to stagger the autumn sowing, it is even better to plant varieties with different maturity classes. That way you can plant at an optimal date to get good vegetative growth and obtain a range of harvest dates through the winter.

There is one more crucial planting season, and that is for onions, garlic, peas and broad beans. In our climate this is late autumn to early winter. Although peas and broad beans like temperatures in the 15–25 degree range, they do poorly when the days get hotter than this, but they will poke along during the cold of winter and can withstand some heavy frosts. The flowers, on the other hand, are killed by frost. So the best option is to plant during early winter and by the time they are big enough to flower, the frosts have gone. Then the pods develop during the warmer spring, and the season is over before the heat of summer.

Understanding the plant response to temperature and day length helps us to work out why some crops fail. If I plant broad beans in late summer, they will germinate quickly and put on a lot of vegetative growth. They may even fall over during a heavy rainstorm, but they will not set flowers until the shorter days of autumn. Flowers will appear in late winter, but these will be killed by the frost before turning into pods. The plant continues to grow and set flowers and by mid-spring the plants are chest high and the later flowers are now forming bean pods. If we had delayed sowing to mid-winter, the crop would flower at a similar time, in response to the day length signal, but would not even be knee high.

Onions are a good example of a crop that needs to be sown at just the right time and stored for the rest of the year. Onions like cool weather during the early stages when the leaves are growing and warm weather to produce the bulb. Plant too early, and they produce a flower but no bulb; too late and the onions will be small.

It is the same with normal peas, snow peas and snap peas. If I plant too early I lose flowers to frost and if I plant too late the plants do not reach full size before flowering. Broad beans and climbing peas are called indeterminate growers, meaning they continue to flower over a long period. This makes them less sensitive to frost and the long harvest periods makes them most suitable for the garden. Large-scale commercial producers prefer determinate varieties. These varieties go through their vegetative stage and switch suddenly to flowering over a short period. The result is that the crop is more or less ready on the same date, which is essential for machine harvesting. For these farmers, the sowing date is absolutely critical.

When I have gardened in a warmer climate with fewer winter frosts, it was better to plant snow peas and snap peas in the early autumn. The crop would grow fast through the cool of autumn and produce flowers and pods before the killing frosts of winter. Winter planting was a bit less successful, perhaps because maturity coincided with weather that was too hot. The point is that if you know how the plant responds to its environment, you can experiment and perfect the sowing window for your location.

The best sowing date for onions can be tricky. The layers of the onion are made up of the thickened bases of the leaves. So the more large leaves on the onion plant, the bigger the onion will be. The earlier you plant, the more leaves you get; however, onions need to experience a period of cold before the leaf base thickens up to form a bulb. If I plant too early, before it gets cold enough, the plants grow well but may decide to send up a flower head without producing a bulb at all. This is called 'bolting' and is extremely annoying. Plant them too late and you get few leaves and a small onion.

I grow my onion seedlings in the glasshouse and plant them out mid-winter. They appear to do nothing for a few months, and then suddenly have a growth spurt through spring, start bulbing in early summer and are picked mid-summer. Onions are one of the crops you need in the kitchen all year round, but you can't grow them all year round. Different varieties do have different responses to day length, which gives the opportunity to spread the harvest, but there are also varieties that will store for a long time without sprouting. I tend to plant three onion varieties around the same time, but with three different longevities in storage.

There are a number of other leafy vegetables that will do the bolting trick. It is particularly common in the lengthening days of spring, where fairly small Chinese greens, rocket, celery and carrots will also send up a flower head instead of doing what we expect.

A number of the common vegetables are not so exacting about their planting dates. They will withstand a moderate frost and not run to seed in response to some day length stimulus. We grow a lot of loose-leaf mixed lettuce and plant them throughout the year, except in the coldest months when they go into the glasshouse. Carrots, spinach and beetroot have a wide sowing window, although there is a distinct effect on quality. Planting carrots in late summer and getting them to mature into the cool weather gives the crunchiest and sweetest carrots.

Carrots are biennial, meaning they complete their lifecycle over two years. In the first year the storage organ is produced; the carrot root itself. In the second year the reserves in the root are drawn upon to produce the flower and finally seed. If the carrot is grown under conditions which are too good, it might try to run straight to seed without forming a storage root. The cool weather of autumn encourages the carrot to store as much sugar in the root as possible, rather than grow new leaves. That is why carrots displaying luxuriant top growth often give a disappointing harvest, whereas those with modest leaf growth have delicious, sweet roots. The available sugar was diverted to storage rather than more growth.

In our climate we can grow a good range of vegetables and supply ourselves for most of the year. Mid-summer to early autumn is the peak picking time for tomatoes, capsicums, eggplants, potatoes, zucchinis, cucumbers, melons, corn, carrots, beans and salad greens. Most of these can be picked up until the last frost, although the plants are often looking a bit ragged by then. We pick the bulk of the potato crop and the butternut pumpkins in autumn and these have to last through the whole winter.

As the cooler weather sets in we are still harvesting lettuce, carrots and spinach and the first of the brassicas. Cabbage, cauliflower, broccoli and kale are the mainstays through winter. Late winter and early spring constitutes the hungry gap. Most of what we planted the previous autumn has now been eaten. Fortunately there is usually something in the glasshouse.

By late spring things are starting to look up. There is some asparagus, broad beans, various salad greens and the pods on the dwarf peas are nearly filled. By early summer we are awash with snap peas and snow peas, and the first of the early warm season crops from the glasshouse – zucchini, cucumber and cherry tomatoes. In mid-summer it is time to pick the onions and garlic, string them up to dry and hope they last till next year. They never do.

Not every crop is a success every year, and that's one of the things that makes vegetable growing so interesting. We make slip-ups in planting dates and get caught by late or early frosts. A heat wave and a hail-storm can lay waste something newly planted or ready for harvest. Pest and disease incidence is very variable from year to year. One year the beans are attacked by green vegetable bugs and in another the broccoli is infested with aphids, but we are not looking for 100% success; 70% will do.

Coming to grips with some of the first principles – the interaction of plants and their environment – gives us some measure of control. Understanding the factors that set the sowing window, and exploiting this to spread the harvest dates, gives the best chance of getting a steady food supply. We shouldn't throw out the sowing date guide, and always be on the lookout for a new variety that can help fill a gap in the season. Knowing the response of each vegetable cultivar to frost, temperature and day length helps to unravel at least some of the mysteries of success and failure in the vegetable garden.

Chapter 3

Intrigue in the orchard

There were no fruit trees in my childhood garden and it never crossed my mind to plant any. The time lag between planting and first harvest was well beyond my time horizon. Years later it didn't seem to make much sense to plant trees in our rental accommodation, but we did. One freezing weekend in mid-winter we dug six gigantic holes in the lawn, filled them with compost and planted two peach trees, a plum, a cherry, an apple and a pear. I only had six months left on my contract at work and about the same length of lease on the house and I wondered who would one day eat the fruit.

The trees grew magnificently that first summer, and even set some fruit, which I quickly removed in obedience to my fruit grower's manual. As it turned out we were still in the same house during the second summer and this time I left some fruit on the trees to taste. By year three we had a substantial harvest of peaches and plums, and the first harvest of pears.

Up to this time a vegetable garden had meant two things; planting summer annuals like tomatoes and zucchini, and winter annuals like broccoli and peas. There were a few longer-lived aberrations like strawberries and rhubarb that kept going for a few years. One year a friend gave me some raspberry and youngberry canes and then we discovered how quickly fruit trees could grow.

So, in my mid-thirties when we finally moved into our own home, the first task was to establish the orchard.

Essentially, plants are either annuals or perennials. The annuals complete their lifecycle within a year, although we generally eat vegetables well before their lifecycle is complete. We eat lettuce and cabbage while they are still in the vegetative stage. We eat the unopened flower buds of broccoli and the young fruit and developing seeds of zucchini and beans. By the time we harvest a tomato or pumpkin the seeds are fully formed, which is why we so often see these two growing out of the compost.

Carrots and onions are biennial, which means they take two seasons to complete their lifecycle. In the first year they make a storage organ and in the second they draw on this storage to produce a strong new stalk with a flower on top. Annual plants race through their developmental stages, trying to get the seed mature before the frost hits or the weather gets too hot and dry. Perennial plants are in it for the long haul, and have a very different way of going about things. First they invest in infrastructure – deep roots, a trunk and branches – before they start to produce any fruit. Secondly there are carryover effects from one season to the other, that makes fruit growing full of intrigue.

The variety of fruit we can grow matches or even exceeds that of the vegetables. There are the berries, like raspberries, boysenberries, youngberries, blueberries and strawberries. Next come vines, with black and white grapes, kiwifruit and passionfruit, and then there is the pome fruit family – apples, pears and nashi. The stone fruit have the most variety – peaches, nectarines, apricots, plums and cherries. We also grow figs and mulberries, and could grow olives and persimmons but have not got around to it.

We are a little cool to grow all the citrus crops, but do produce a lot of lemons and mandarins. We cannot grow any of the warm climate fruit, like mango, pineapple, banana, pawpaw or guava, although we have planted avocados. There is a huge variety of lesser-known fruits of which we have tried a few, like feijoas, tamarillos and strawberry guava. We do not grow any of the nut crops, because we are not warm enough for most of them, nor have the space.

At last count there were 35 fruit trees in the garden, 15 vines and several rows of berry crops. They would take over the whole garden were it not for a strict trellising and pruning regime. Our block is slightly wedge-shaped, with a narrow front and a larger back. The easterly-facing front fence line is espaliered with plums, apricots and peaches. The espalier continues along the northern boundary carrying apples and pears and then turns into a T-shaped trellis for vines that lifts them above fence height. Pergolas along the western boundary support kiwifruit and netted areas for more stone and pome fruit, particularly cherries. Strawberries, raspberries and boysenberries grow alongside paths, so they can be snacked on by passers-by.

I try to have some ripe fruit to pick on each day of the year. For about five months through summer and autumn there is more fruit on offer than we can possible eat. In winter we are down to citrus and in spring there is not much on offer.

Whereas the young tender vegetable crops struggle through the cool days of the new season, the fruit trees burst into spring. Apart from the citrus and feijoas which hold their leaves all year round, the trees are bare branches and twigs in the bleak days of winter. By early spring they are covered with blossom, from the brilliant white of the cherry through the pink-white apricot and pear blossom to the bright pink of the peaches. A month later the blossom has gone but the new shoots have already made a dense canopy of green. This is the great advantage of being a perennial. The root infrastructure is in place, the buds are already formed and ready to be powered by the sugars and nutrients stored in the woody tissue from the year before. Once spring hits, the growth is magnificent.

Fruit trees, like all other plants, have to partition their energy between growing roots and growing shoots. Obviously there has to be a balance here, because water is supplied via the root system and lost from the leaves. Trees must also balance the partitioning of energy between leaf growth and fruit growth. The tree has a number of hormones which keep some sort of a balance between the below-ground and above-ground parts and the leaf growth and fruit growth.

New roots have already begun to grow in the cold spring soil before the leaf buds burst in spring. Such rapid leaf growth will need a strong supply of water and nutrients, so the root system needs to get a head start. If the roots can't grow vigorously in early spring, because the soil is dry or hard, salty or acid, then the shoots will similarly lack vigour.

Trees make spectacular regrowth after a hard pruning because the tree puts all its resources into shoot growth to get the above and below ground parts back into proportion. Similarly, if the tree roots get killed by temporary waterlogging or are pruned by soil cultivation, the top growth slows down until the roots have grown back again.

The young tree has an upright growth form, trying to get as much height as possible so it does not have to live in someone else's shade. This upright growth habit is due to a hormone produced within the growing tip of a shoot. The hormone slowly moves down the branch, inhibiting any side shoots from growing. With the competition out of the way, the topmost shoot has all the resources for growing, which is why the vertical growing branches grow so much faster than the others.

Left to their own devices, most fruit trees will get much bigger than we want them, and since these vertical branches rarely produce fruit, we cut off the growing tips and bend the branches on trellises. Removing the growing tip removes the source of the side-shoot-inhibiting hormone, but bending the branch can have the same effect. The hormone does not move down a near horizontal branch, or even one at a 45 degree angle, so the side shoots escape its effect and we get the side branches we want.

Some trees have the frustrating habit of biennial bearing, which means they produce loads of fruit one year and hardly any the next. This is the work of another hormone that is produced in the seed of the very young fruit in spring. When a young peach shoot grows out in spring, primordial buds form at the base of each leaf. These are the buds that will shoot out 10 months later in the following spring.

In late spring, the tree has not decided whether these newly forming buds will become shoot buds or fruit buds. This is determined by the load of fruit it is carrying. If, by early summer, the tree is covered in tiny peaches, a large amount of hormone will be produced and the developing buds will turn into shoot buds. The next spring the tree is all green, with no blossom. Conversely, if the spring tree carries almost no fruit, then the developing buds will mostly be turned into fruiting buds for the next year.

We call this balance between shoot and fruit growth a source–sink relationship. Leaves are the source of sugar and fruit are the sink for it. If the sink is too big, the fruit will be numerous, small and poorly flavoured. If the sink is too small – too few fruit – then the tree just gets bigger and bigger as the resources go to growing branches and leaves. The hormone in the developing fruit is one of the ways the plant keeps the source and the sink in balance.

Biennial bearing is the process gone wrong, where the tree swings between the two extremes. The solution is to remove many of the excess young fruit early in summer of the 'fruit' year. The supply of fruit-bud-inhibiting hormone is reduced, and so the ratio of fruit to shoot buds is restored for the following season. Even trees that are not extreme biennial bearers still tend to show a pattern of slightly up followed by slightly down years. It is a reminder that the business of fruit bearing is at least a two-year process. In the first year the bud is laid down as fruiting or vegetative and in the second these buds grow out into fruit or shoots.

A tree covered in blossom is no guarantee that there will be a tree laden with ripe fruit. There are lots of things still to go wrong and the most feared is frost. The pome and the stone fruit want a cool winter, during which they go dormant, but a spring frost is lethal to blossom and young fruit. Thus it is essential for a perennial crop to go dormant and then flower at the right time, and this happens in a remarkable way.

Each type of tree has what is called a 'chilling requirement' which we define as the number of hours below 10°C. Canberra has more than 1300 hours of the year below 10°C, which is sufficient for all stone and pome varieties. Coastal cities at similar latitudes have half the hours below 10°C, so many varieties do not go dormant and do not flower at the right time there.

Cherries have quite a high chilling requirement, between 800 to 1200 hours depending on the variety. After these hours have been counted off, the tree then responds to the number of hours above 10°C. These are called 'heat units' and when the requirement is satisfied the buds start to open. Elaborate as this timing system is, rogue late frost still wreaks havoc.

There are many other intriguing facets about fruit growing. For example most fruit trees are grafted. The root half of the tree is a different variety – and sometimes even a different species – from the top half. There are many reasons for grafting. Some rootstocks are resistant to a particular disease, while others are tolerant to adverse growing conditions like salty soil. The rootstock is also used to control the size of the tree. There is a whole range of apple rootstocks that give different size trees when twigs from the same apple tree are grafted on top; another remarkable example of how parts of the tree communicate with each other.

The most important reason for grafting is because each fruit tree variety is an exact clone of some special chance mutation. One of the most celebrated is the Granny Smith apple which appeared from the seeds growing from discarded trees near a creek on a little farm which is now the Sydney suburb of Ryde. If you get the seeds out of a bean or tomato and sow them next year, the new plants will taste the same as their parents. Not so with the apple. If you sow the seeds of a Granny Smith apple you will most likely get a green, rather tart apple, but it will not have as good a flavour as the original. The arrangement of genes in the Granny Smith is so subtle that sexual recombination, as occurs in the flower of the tomato or bean, scrambles the genes sufficiently to lose the taste we are after. Every Granny Smith apple tree in the world is a clone, a graft originating from that chance seedling that grew in a Sydney garden in the late 1800s.

In our garden, plums, grapes and kiwifruit produce the most fruit with the least effort. Those who rely on shop bought fruit will likely have missed the wonder of the humble plum. We grow several different varieties each with their own superb and distinct flavour.

Most of the varieties we know so well had this inauspicious start. Oranges originated in Asia, but it was a missionary in Brazil who noticed a single tree produced fruit with no seeds. Without seed, there was no way the tree could reproduce itself, so the missionary took some cuttings and sent some to California. All navel oranges in the world today are clones from those original cuttings.

Pruning is pretty much essential for most of the fruit we grow. The most baffling part of any fruit growing book is the section on pruning because each type of tree needs different treatment. I study the before and after pictures, but when I go outside I find it near impossible to put into practice what I have tried to memorise. It is best to try and understand a few principles.

We prune to keep the tree a manageable size, and to thin out the branches growing in the wrong directions so we get the shape we want. A small 'open' tree usually has fewer and larger fruit, is easier to net from the birds, and better air circulation helps with disease. But the main reason for pruning, and the key to getting it right, is to favour the 'type of wood' which produces fruit. To get this right, you need to watch how the tree grows.

Peach blossom is only produced on one-year-old wood. The previous spring a shoot bud will grow, say, 50 cm long with 20 leaves and 20 buds. In autumn the leaves fall off, leaving the 50 cm twig and 20 buds. The next spring, we call this a one-year-old branch. Some of these buds will produce blossom and others leaves. Older wood will not produce any blossom, so the idea is to cut away some of the older wood, leave sufficient one-year-old wood for next season's fruit, and some buds to become the one-year-old wood that will produce fruit the year after.

Apples and pears are different. Most apple and pear blossom does not grow on one-year-old wood but on older wood called spurs. These look like small compressed branches. And to keep us on our toes, grapes produce flowers on current season growth. That is why we can prune a grape to almost nothing. The new shoot bursts out in spring and the flower cluster forms near the base of each new shoot. It sounds complicated and it is at first. You won't kill a tree by pruning, so it best to have a go and then watch carefully what happens.

Our fruit year starts with the harvest of strawberries in late spring. The cherries start in early summer and if the weather is dry we get a spectacular harvest. If it's wet, they split and rot a lot faster than we can eat them. By summer the apricots are in full swing, as well as the raspberries, blueberries and boysenberries. We get the first peaches around Christmas, and used to harvest different varieties of peaches and nectarines right through until mid-autumn; however, we had so many pest problems with the later fruiting varieties that we dug them out.

Grapes from the glasshouse start in mid-summer and different varieties take us through to late autumn. We can also add plums to the list: blood plums, greengages and the delicious purple 'Quetsches' plums from mid-summer onwards. In late summer the first apples are ripe, followed by nashis and pears. The raspberries give a second crop in autumn. The last fruit to come in are the aromatic feijoas, and the kiwifruit, which keep in the fridge for weeks. Through the winter we get mandarins, followed by a lull as we wait for the whole cycle to begin again in spring.

Some fruit trees seem to produce well with almost no care. Our blood plums are laden with fruit year after year with no special treatment. Most of the others are up and down, for with fruit growing there is a lot that can go wrong. At the end of each season we vote on the tree that performed the best, and the tree that made the best improvement on the year before. The best performing tree is usually won by one of the plums, grapes or cherries. This year a peach took out the prize, with most improved going to the figs.

Chapter 4

Feathered thieves

Biodiversity is a buzzword these days. The word is a contraction of the phrase the 'diversity of biological life'. The great diversity of life is a reflection of the diversity of the environment. Climates range across the continents from hot to cold and from wet to dry. At the next scale down there is the variation due to altitude, with the higher country being moister and cooler.

Soils differ from rich to infertile, depending on the parent material, vegetation and climate. Wind and water erosion strip some areas of their soil, while others, like river valleys, accumulate sediment and nutrients. Fires, extreme weather events and disease outbreaks also contribute to the heterogeneity of the landscape, and so to the diversity of life.

Since we need to be clothed and fed, we harness the natural resources to meet our needs, using domesticated plants and animals. Agricultural landscapes have a beauty of their own, but they are more homogeneous than the landscapes they replaced, and this has far-reaching consequences for biodiversity.

The natural vegetation around my city of Canberra is open eucalyptus woodlands interspersed with tussocky grasslands. It is not very good farming country, but has proved a good region for fine wool merinos. Trees have been cleared from most of the flat country over the last 150 years, and replaced with introduced pastures. These days there is a movement to replant some of the trees for a range of reasons, including tackling salinity, soil conservation, carbon sequestration, increasing biodiversity and diversifying farm income.

An organisation called Greening Australia has been incredibly effective in mobilising farmers and volunteers to plant native trees and shrubs. Significant revegetation has taken place on 400 properties in our region, and Greening Australia wondered what impact this would be having on biodiversity. Surveying the change in bird populations was one way to get a grip on this, so they asked the Canberra Ornithologists Group to help.

This group of keen birdwatchers had already done some pioneering work in the nation's capital, recording a mind-boggling 216 different bird species from observation points across the city. We call Canberra the 'bush capital' because there is a large amount of Nature Park and open spaces spread through the city. There are hills and ridges around the centre of town, and open areas connect these across suburbs to the surrounding countryside.

Presumably each bird species has slightly different food, breeding and other requirements essential for survival. Such a vast number of species must tell us something about the diversity of the area, even if it is not immediately obvious to the untrained ecologist. If some species are becoming more common and others less so, then this may be telling us something about our changing environment. This is not the 'canary down the mine' idea, where the bird keels over if the carbon monoxide levels get too high. Changes in bird populations could be telling us about subtle interactions taking place that we could never hope to unravel from first principles alone.

The Canberra Ornithologists Group monitors bird sightings at over 270 sites around the city and they have done so for over 20 years. This is all carried out by volunteers, who record the birds they saw in a given week within 100 m of their chosen observation point. Some sites may be monitored just one week a year and others over 40 weeks per year. A few sites have been monitored every year and some for just a few. Nevertheless, when it has all been put together we get a fascinating picture of what is happening in the city.

Many species are extraordinarily rare, with 62 species showing up five times or less in over 1000 weekly records. Another 119 are seen in less than 10% of the records. A keen volunteer who records bird sightings over 10 weeks of the year would be lucky to see these species just once. At the other end of the scale are those birds we see all the time. There are 15 species that are reported in more than nine out of 10 of the weekly reports. These are all the usual suspects: magpies, cockatoos, galahs, parrots and the like. No doubt this group benefits from human activities.

The extent to which the rare species are inherently rare, or have been pushed there by habitat change, is hard to determine. Certainly there are a number of species that have become locally extinct, or exist in sub-viable populations because their habitats have been drastically reduced. Some species seem to persist at quite low levels while other species show an amazing ability to increase in numbers when conditions are favourable.

In the 2 million hectares of African savannah covered by Kruger National Park, there are over 517 bird species represented. In this wilderness, 80% of the estimated bird biomass comprises fewer than 2% of the species. This is not skewed by the hefty ostrich or the flocks of vultures, which each account for around 1% of the total biomass. Of the estimated 40 million birds in the park, 80% are red-billed quelea. The small finch-like birds add up to more than half of the total weight of all birds. The vast majority of species are naturally found at low densities.

Greening Australia's take on biodiversity was to focus on species known to be declining in numbers, because these species were probably most affected by the change in habitat. It turned out that the 'decliners' were showing up in their surveys where they had replanted trees. By planting large enough areas, and enhancing and connecting existing remnants, it was possible to bring some of the decliners back. As we might expect, the number of different species observed increased with the age and the width of the revegetated area because some birds have very specific requirements.

Thornbills, robins and babblers were among the species that were seen to be increasing in revegetated areas. The tiny thornbill requires a hectare or two as a breeding territory; a pair of robins wants over 10 hectares comprised of trees for nesting and cover and open ground for feeding; the white-browed babbler must disperse several kilometres to find new territory, but will not travel more than 20 m from shrub cover.

For those of us who eat bread, we can't expect the farmer to give up a hundred hectares of arable land for a few small birds with big empires. Obviously there will have to be some compromise. So what does this compromise look like in the home garden?

A long time ago, someone cut down the few eucalyptus trees that inhabited our garden, and I have planted more than 50

Growing fruit is lots of fun but a lot can go wrong. First the tree needs to set plenty of fruit buds (blossom). These need to pollinated, escape the frost, and in the case of apricots escape the crimson rosellas which strip blossom off the trees. The scarecrow cat is on guard against the rosellas.

species of edible plants in their place. The environment is now totally different, which has worked against a number of the original inhabitants and favoured others.

The currawong is garden enemy number one. These feathered thieves are stealthy, ubiquitous and never miss an opportunity. Mulberries and strawberries are the first fruit of the season and currawongs get every one that is not protected. Then they move on to the cherries. I have one tree in a position that is difficult to net. It has a very thick canopy and the young green fruit are almost invisible. I usually wait until the fruit starts to go pink and the birds start to take notice before netting. In the last couple of years the currawongs have beaten me to it. I leave for work with hundreds of perfect ripening fruit on the tree and return that evening to find none. Not one.

During winter the currawongs get more desperate. They will destroy a bed of neatly transplanted onions, for what purpose I'm not sure. They get more devious each year. Lately I have seen them dive-bombing the nets draped around the cherries and grapevines. The impact of one heavy bird pushes the net against the branches and the others reach through and pluck the booty with their long beaks. If there is a hole, a currawong will always find it. I have picked grapes and put the container down for a short time to attend to another task. As soon as my back is turned, the currawongs attack. I have come to understand that there is always a currawong watching.

I have started to fear the huge flock of sulphur-crested cockatoos that roosts in the gum trees in the park at the end of the street. They occasionally venture from the park to the nature strip, but never into the garden. Until recently. Last year they laid waste a few fruit trees between sunrise and the time I got out of bed. The ground was littered with fruit, each with just one bite removed.

Crimson rosellas eat the fruit too, particularly apples. Apart from their frustrating habit of stripping apricot blossom, it's hard to hold anything against these spectacular birds. They also tend to eat one apple at a time, and although we see them on a daily basis, I am sure it's the same local pair, so there is a limit to how much they can eat. There seems to be no limit to the damage a flock of cockatoos can do, and the bird survey suggests their numbers have increased threefold in 20 years across the city.

We have our fair share of the introduced quartet of menaces: mynas, starlings, sparrows and blackbirds. All are common and unwelcome. We have seen less of the first three, since the chickens were cleaned out by a fox, so I suspect they thrived off chicken food and food scraps thrown into the pen. There has been a concerted effort to trap and dispose of mynas in the city, because they tend to muscle other native birds out of the area, and their numbers have gone down. Our new chicken feeding arrangements seems to have outwitted the starlings and sparrows for now.

Then there are the faithful garden friends. The bird that probably puts in the longest hours in our garden is the magpie-lark or 'peewee'. A pair of peewees is almost always on duty, patrolling the rows and stopping frequently to pick at this or that. The magpies, big brother of the peewees, do most of the heavy lifting, listening intently to sounds below ground and digging out the offenders with their sharp beaks. The magpie staple seems to be the larvae of scarab beetles, and the scarab is a real pest in our garden.

Scarab larvae turn into brown beetles around Christmas time, hence their common name of Christmas beetle. They are strong flyers with slow reflexes and crash around lights on warm summer nights. The scarab eats gum leaves and is partly responsible for dieback of eucalypts across huge areas of farmland. The beetle larvae hatch underground, and we tend to only notice them when they have grown to a fat white grub curled into a 'C' position. By this time they have chomped their way through plenty of roots.

Scarab larvae can completely separate grass from its root system, to the point that it becomes dangerous on sports ovals if someone slips on unattached turf. I find scarabs under almost all crops in the garden, few in some years and plague proportions in others, and always under a discarded piece of wood or brick. Maybe they prefer the damper soil under these obstacles, but my guess is they are protected there from magpies. I think I would be in trouble without magpies.

Silvereyes flit through the branches in groups of 10 or 20, never staying in the same place for more than a few seconds. I have watched them feed in trees laden with aphids, but they never stop to finish them all, preferring to keep moving and return the next day. In winter they are joined by wrens and thornbills. The latter feed near the ground, where there are even richer pickings among the vegetables. Yet the wrens and thornbills disappear in spring and summer when aphid numbers are rising and the silvereyes rarely descend from the trees. I often see zucchinis or beans thick with aphids that seem to be a missed opportunity.

Willie wagtails and fantails specialise in flying insects. The wagtails usually launch their attacks from the ground and the fantails from the branches. I would like to see much more of these two, but they also push off in spring. Perhaps there are better pickings in the surrounding bush. In late summer and autumn clouds of cabbage white butterflies lay eggs on the brassicas with impunity. Even I can catch them by hand, so I'm bamboozled as to why these aerial acrobats don't seem to bother.

Another two pests that gets off lightly are slugs and snails. The French eat them, so why not the birds? They mostly feed by cover of darkness, but rain after a dry spell brings them out in droves at twilight. Even a family of white-wing choughs, who can completely devastate a neatly mulched bed with their foraging, don't seem to make an impact. It seems like a big

I love the diversity of birds in my home city, but unfortunately a few are a menace to the urban farmer. Sometimes I think I would like to throw a net over the whole garden and keep them all out, but I know this would do more harm than good.

missed opportunity, a vast feast waiting to be discovered. Perhaps the widespread use of slug pellets, whose active ingredient is also used as a bird repellent, has warned them off.

For a while I kept some fruit trees in the glasshouse to give them 100% protection from the enemies outside. In the first year I got terrible fruit set. The insects that do the pollinating had also been excluded. The following year I laboriously pollinated in the glasshouse by hand using a paint brush. This time the trees were laden with fruit, but later the leaves turned a dull mottled yellow. I had a major mite infestation – tiny spider-like creatures almost invisible to the naked eye sucking the juice out of the leaves. Where there were no mites there were aphids. Outside the glasshouse, just a few metres away, the same trees flourished unharmed.

The experience made me ponder that curious term, 'ecosystem services'. Ecologists use this term to describe the things we get for free from the biodiversity around us, but do not really realise until they are gone. Pollination is an obvious example. There are natural processes that clean up air and water, and there are the birds that predate on some of the 86 000 species of insects out there. The lesson is that by simplifying our environment we make ourselves more vulnerable, not less. I barricaded my trees from the enemies I knew, only to lose the friends I did not know.

At this point we face a dilemma. The red-billed quelea move over large distances in huge flocks and will breed at short notice when conditions are favourable. Males can build a nest within a day and hundreds of thousands of females can lay their eggs within a few days of each other. The incubation and nestling period are among the shortest of any bird, and within a month the colony can double in size to up to a million individuals. Onlookers say the savannah looks like it's on fire when a flock this size moves. Such numbers of quelea can be devastating to the small-scale farmers trying to grow enough cereal to feed their families.

It is easy to start to categorise the species as friends or foes. I like the rosellas because they are so beautiful, or the peewees because they are so obviously useful, but I don't like the thieving currawongs. In this case biodiversity is evaluated according to my personal taste. Then we have a soft spot for the birds that we do not see often in the garden. We never want a bird to go extinct, or at least locally extinct, so we will always want to conserve something which is rare. Some people promote the idea that every species has a particular function, and if we lose that species then the system is somehow more vulnerable to collapse. We never want to lose a species, but I have difficulty with this particular line of argument.

The managers of Kruger National Park are charged with the conservation of biodiversity. For many years they focused on the iconic species, like the rare antelope. After all, tourists come with a checklist, and they want to see as many animals as possible. After many years of research, the scientists in charge have given up on the idea of focusing on individual species. Every time they tried to optimise for one specific outcome, something else went wrong.

Now they manage for the heterogeneity of the bushveld as a whole. Heavy grazing by elephants, natural or introduced fire, provision of water or management of disease can each contribute to the homogenisation of the veld or contribute to its variability. Variability gives rise to many different habitats, so managing for heterogeneity gives the best chance of conserving biodiversity as a whole.

Agriculture, of course, needs long straight rows of uniform crops that can be planted and harvested on a large scale. It seems an impossible task to conserve biodiversity in all but a few areas protected as national parks. Elephants and lions will never be compatible with agriculture, but there is a strong case to retain and enhance the natural biodiversity of agricultural landscapes. A major study in Australia showed that biodiversity could be largely conserved when one-third of the land was under intensive cultivation of crops, one-third under low intensive use, such as occasional grazing, and one-third left in its natural state. It is an inspirational target to aim for.

Chapter 5

The lapsed organic gardener

I am a lapsed organic gardener. I used to run an organic garden, but for various reasons, changed direction. I aim to get back to a more organic approach, but there are some formidable obstacles in the way.

Very early in my gardening career, I got a soil-testing kit as a present. It contained test tubes, reagents and colour charts. I had to put some soil in the test tube, add water plus the reagents and shake it all up. After settling, the liquid above the soil had turned a yellowy or bluey colour, depending on the nutrient I was measuring. The intensity of the colour was determined by matching against a chart, and this told me the level of nitrogen, phosphorus and potassium in my soil.

The kit had instructions on how much of each fertiliser was needed for my garden, based on the colour charts. I went off to the garden shop and bought small bags of ammonium nitrate, potassium sulphate and superphosphate, which I weighed out on the kitchen scales. It all felt very scientific for a gardener not yet in his teens.

In those days I gardened on very sandy soils, and fertiliser made a big impact. Yet I soon noticed that the tomato or pumpkin that self-seeded out of the compost heap grew faster and more luxuriantly than anything in the garden. I started to read books on organic gardening, and found out that compost was the cure for everything.

If the soil was too sandy, organic glues in the compost helped to stick the sand grains together to form soil crumbs. If the soil was too heavy, the same glues held the clay particles apart, so the air, water and roots could penetrate. As the organic matter decomposed, it released nutrients at the rate the plants needed them. I did not have to worry about measuring out an N:P:K formula (the ratio of nitrogen, phosphorus and potassium in a given fertiliser) or about a deficiency in some trace element. Compost was formed from plant material and therefore contained all the elements a plant needed, and in the right proportions.

Compost helped the soil to hold more water and to drain faster when the soil was too wet. It helped to keep the pH of the soil in the range the plants liked. Compost even provided protection against pests and diseases. I lived in fear of the tiny nematode worm that attacked the roots of tomatoes, but certain types of fungi in the compost have filamentous strands that loop around the unsuspecting worms and tighten like a lasso.

Best of all, compost was built from the waste products of the garden. Dead plants, leaves, grass and hedge clippings, and the spent straw from the chickens all went into producing this delightful sweet-smelling stuff that made everything grow. Compost is nature's perfect cycle of growth, decay and regeneration. I became a dedicated organic gardener because it made sense, and because it worked.

Modern organic farming goes well beyond just using compost. Fundamentally it is about seeing the farm as a 'living organism' and that a 'healthy living soil' translates into healthy plants, farm animals and people. Organic principles also encompass fairness in the food production, processing and marketing chain, including the welfare of animals.

Organic farmers aim to reduce reliance on fossil fuels, recycle nutrients locally, and they seek to understand and enhance the natural processes that keep soils fertile and pests in check. Why use fertilisers mined far away or produced from energy-hungry industrial plants when it can be recycled for free? Why spray for pests when their natural predators will keep them under control? Why use herbicides when crop rotations can suppress the weeds?

These questions were answered the day I started my first full-time job. Where do I get enough material to make compost for several hectares of lettuce? Some farmers do of course apply compost to large areas, but it's not simple, at least not as simple as the fertiliser injection pump that dosed precise amounts of chemical nutrients into irrigation water each day and delivered them uniformly through thousands of drip emitters to each little plant.

My job was to grow the vegetables for a large field experiment which was to demonstrate the concept of solar-powered irrigation. A lot of money had been sunk into the venture, and since it was a showcase experiment, a good number of dignitaries came through. The crops had to make an impression, with minimal pest and disease problems, and they would be sent to market as part of the overall economic analysis. My university training had not prepared me for this, so I made friends with local market gardeners, read their magazines and took their advice.

There was an advert that regularly took up the whole page of our grower magazine. Prominently displayed was the picture of a fat caterpillar with the words, 'Guess who's eating at your place tonight?' above it. I often lay in bed at night pondering that question.

When I grew lettuce in my vegetable garden, I knew they had some enemies. Slugs and snails were always a menace and plants tended to rot a bit during wet weather, sometimes going slimy inside. Occasionally the odd plant never seemed to get going and remained stunted. What I didn't know was the formidable array of insects and viral, bacterial and fungal diseases that were waiting to undermine my first attempt at being a real farmer.

There is a disease of lettuce called necrotic yellow virus. A lettuce that gets this virus grows very slowly, fails to heart and turns a kind of fluorescent yellow. The virus lives in weeds, particularly the common milk thistle, and is transmitted to the lettuce by an aphid. This aphid ingests the virus when sucking the sap of the thistle and injects the virus when they move to feed on lettuce. It is not possible to know if the aphids are carrying the virus, and it takes some time to see if the lettuce has contracted necrotic yellows.

Moreover the virus is injected in the first feed, so you can't wait around and see what happens. There needs to be a zero tolerance policy towards aphids, unless you want to have your field sprinkled with conspicuously diseased plants. The remedy was to apply a systemic pesticide to the lettuce. Systemic means that the lettuce itself takes up the poison, so the feeding aphid dies at its first meal.

Then there were the fungi. A spell of wet weather followed by some warm humid days could mean disaster. We needed to get preventative sprays onto the crop before the rain, because by the time the diseases took hold it was too late to stop them. Meanwhile the weeds were growing as fast as I could hoe them out. The agricultural chemicals that I had so despised were becoming allies in the fight to grow the perfect crop.

I continued to grow lettuce without these chemicals in my home garden. It did not matter if the plants did not all mature at the same time, or if a few leaves were eaten or diseased. Our first Canberra garden was on a large inner city block. Up to half the garden grew clover and lucerne, which we cut and used for mulches on the vegetables. Both these plants are legumes, which can turn nitrogen gas from the air into nitrogen fertiliser for plants. Lucerne also has wonderfully deep roots which can scavenge nutrients from deep in the subsoil and return them to the surface.

Then we moved to a more normal-sized block, the family and the house got bigger, and the garden got smaller. The clover and lucerne beds were slowly replaced by potatoes and

These farmers are applying fertiliser to their tomatoes individually to each plant using a teaspoon. Despite their material poverty, they consider it money well spent.

peaches and all the other things we could eat ourselves. We had to become more productive, and we had to find new ways to replace the nutrients that were exported down the sewer or leached below the shallow roots of our crops, because we did not have space for cover crops.

We made as much compost as we could, purchased manure, used blood and bone and wood ashes. A lot of nutrient was also imported via the chicken feed. It solved our problem, but we were still removing nutrients from some other place, so in one sense we had just shifted the problem. How were the people who supplied us with nutrients, albeit in organic form as manure or chicken feed, returning it to their properties?

We had experienced in microcosm what the farming world had already discovered. When you have to produce more food from less land you have to change the way you do things. If you take a lot of nutrients out of the soil and send them to the city, you have to replace them somehow.

The nutrient that has had the major impact on world agriculture is nitrogen. Nitrogen is an essential part of proteins, and the most abundant protein on earth is the enzyme involved in photosynthesis. This enzyme, called Rubisco, turns carbon dioxide from the air into a simple sugar, and is packed into the leaves of every plant.

There is plenty of nitrogen about – the air is 78% nitrogen and 21% oxygen – but nitrogen gas cannot be used by plants. Plants can only use nitrogen when it is in a different form – as nitrate or ammonium – and it takes energy to turn nitrogen gas into the form that plants can use.

Lightning is one way this happens. Some nitrogen gas is converted to ammonium and rains down to earth during electrical storms. But the most important source of nitrogen fertiliser comes from a remarkable cooperation between a group of soil bacteria and a class of plants known as legumes. The Rhizobium bacteria have the biochemical machinery to turn nitrogen from the air into fertiliser nitrogen. To do this, they need a supply of energy, and the legume is happy to assist. The bacteria infect the root of the legume, forming a little gall or nodule. There they tap into the sugar supply provided by the legume roots, and give nitrogen fertiliser in return.

In the early part of the 20th century, two German scientists worked out how to convert nitrogen from the air to fertiliser nitrogen on an industrial scale, for which they were awarded a Nobel Prize. The industrial production of nitrogen fertiliser is now so important that the world would starve without it. The population of the world doubled in the 40 years between 1960 and 2000. The use of nitrogen fertiliser increased eight times during the same period.

Nitrogen fertiliser and improved crop varieties may feed the world, but there is a downside. Excess nitrate gets washed into waterways and contributes to explosions of algal growth which suffocates other aquatic life. Nitrate also accumulates in ground water, so the water is no longer fit for drinking. Other forms of nitrogen escape to the atmosphere, where they degrade ozone and act as potent greenhouse gases. Worst of all, nitrogen fertiliser is also very demanding of energy. The greatest single energy input into agriculture is through the factories that make nitrogen fertiliser, and energy is getting very expensive.

Phosphorus is a major plant nutrient that is naturally deficient in many of the world's soils. Most phosphorus is mined as phosphate rock, ground up and spread over fields. As a natural product, this is permitted in organic agriculture; however, phosphate rock is almost insoluble, so it is treated with sulphuric acid to make the fertiliser called superphosphate. Mined phosphorus is a non-renewable resource, and supplies are running low.

After fertilisers, the most important chemical inputs into agriculture are the herbicides to kill weeds. The organic farmer keeps weeds under control by rotating crops and pastures, so no one weed can ever get a stranglehold. Weeds are also killed during fallows, land preparation and regularly hoed out during the cropping period. As farms get bigger and specialise on fewer crops, this becomes increasingly difficult. Regular cultivation of the soil also has its problems. Soil crumbs are broken down to dust by repeated ploughing, and fallow fields are vulnerable to soil erosion without the protection of crop residues.

Herbicides ushered in an unexpected revolution. If weeds could be killed without cultivation, then the straw from the previous crops could be left on the surface to protect the soil. The next crop could be sown as soon as the new season rains started. Sowing equipment was developed that could plant seeds through the decomposing straw. A system of minimum tillage agriculture was born. This is a bit of a dilemma for the anti-herbicide camp. Is it better to burn the crop stubble, repeatedly cultivate the soil and leave the land exposed to erosion, or kill the weeds chemically and keep the soil covered with decomposing straw?

I recently visited one of the biggest carrot growers in the southern hemisphere. As always when I visit such farms, I am astounded by the sheer scale of the operation: centre pivot irrigation systems watering vast circles of crop. Harvesters crawl down the rows, lifting a thousand carrots a minute. These are hauled in huge bins to the washing and packing sheds

The hardest thing to get used to when moving from a backyard garden to commercial agriculture is the scale of operation. It's one thing to go into the garden and pick half a dozen carrots for the evening meal. It's quite another to pick, wash, size, sort and package thousands of carrots a minute.

where they are washed and brushed in contraptions like giant car washing machines. Then they are sized and sorted and packed with a hundred pairs of hands into plastic bags, before being loaded into cooled trucks for the supermarket.

I asked the farmer, 'what do you do after you've learnt how to grow 100 tons of carrots a day?' to which he replied, 'learn to grow them organically'.

I assumed this was because the supermarkets, which had increasing power over the whole production chain, were pressuring the farmers to become organic, in line with a shift in public sentiment. The farmer was dismissive about this. He considered the public demand for organics to be just another symptom of how an increasingly urbanised population had become disconnected with the realities of how most of their food was grown. So why did he aspire to more organic methods?

Despite the tractors and implements to loosen up the soil, the irrigation systems to water the plants, the fertilisers to feed the crops and the other chemical needed to keep weeds, pests and diseases under general control, yields showed signs of decline. He could control most things, he thought, except the soil biology, and he had come to the view that somehow the soil biology held the secret to productivity in the long term.

This farmer was reading books published over 50 years ago, trying to rediscover clues to sustainable agriculture. He had started to grow crops like mustard that are known to produce substances that kill soil-borne diseases, and bulky crops, especially the tall forage grasses, which were ploughed back into the soil 'just to feed the bugs' before planting the carrots. He said he was prepared to learn how to farm all over again.

I see the same thing in a backyard. Our current home garden was covered in bushy shrubs when we moved in. We used chains to winch them out roots and all. The growth of vegetables in the first couple of years was phenomenal in this new ground. Looking back at old photos, we do not seem to be able to match this luxuriant growth, no matter how carefully we manage the soil water and nutrients. There is something wonderfully mysterious about virgin ground.

In a way, all farming was organic up to the 19th century because there were no artificial fertilisers. Around the middle of the 20th century, fertilisers and pesticides became more readily available, and the era of industrialised agriculture was born. Organic farming is a reaction against industrialised farming because of a growing perception that the environmental footprint of agriculture is too high.

Australia leads the world in terms of proportion of land area under organic management – a massive 38% – but this is heavily skewed by vast low-input grazing systems on the northern savannah. These systems can be classified as organic because hardly any inputs are used at all. Only around 3% of the total food produced in Australia is produced organically. There are a number of countries in Europe which have 5–15% of their agricultural land under organic management, demonstrating that organic farming is already a viable alternative. Less than 1% of the world's agricultural land is currently farmed under the organic label, although many resource-poor farmers by default use no chemicals.

Perhaps we are still en route to a viable form of organic farming. Chinese agriculture is synonymous with the idea of millions of peasant farmers returning their waste products to the land and retaining its fertility over the eons. But with 23% of the world's population on 8% of the world's agricultural land, and with mass migration to the cities, China has had to embrace the full Green Revolution package. Agricultural production has increased steadily. Now, because of the cost of the agricultural inputs, there is a shift back to organic farming. The area of land farmed organically in China increased more than ten-fold between 2000 and 2006.

Organic farming, as an enforced set of practices, is a bit too restrictive for me. I understand why there is a ban against synthetic chemicals, because the aim is to farm within the limits of the natural system. But I'm not sure why it is more natural to plough up the weeds under the power of fossil fuels than it is to use a fairly benign herbicide. I have stood in the fields of African subsistence farmers and watched the pests devour their only source of food. I have seen countless under-nourished crops, which will translate into under-nourished kids before the next harvest comes in.

Industrial agriculture is currently feeding the world. But like my farmer friend who grows the carrots, there is a whole lot we do not understand about producing food in the long term without undermining the very resource base on which it depends. Our organic farming friends still give us much food for thought.

Chapter 6

Outsmarting pests and disease

A friend I stay with when working in Africa lives on a small farm growing nuts and grazing sheep. Monkeys got most of the nuts last year, and baboons descended from the rocky outcrops around the farm to steal the remainder, breaking large branches in the process. Flocks of guinea fowl scratch out the newly planted oats, sown to fatten lambs. Some as yet unidentified beasts are digging up tracts of land at the back of the orchard, while the ever-present termites bore their way through the trunks of the trees.

Jackals prowl the animal pens at night. Recently a half-grown bullock was killed, partly eaten and dragged some distance, and we thumbed through a gruesome book of pictures trying to figure out which large carnivore could have done it. The farm is nestled in a valley surrounded by native bushveld, and it seems the resident biodiversity is keen on moving back in.

Back in suburban Canberra, the brush-tailed possums dine well on our fruit trees in summer and the vegetables in winter. The density of possums around us must surely be much higher than in the surrounding bush, courtesy of the suburban gardens. But mostly when we think of agricultural pests we are thinking of the small ones; the insects and diseases from around the world that follow the crops we plant.

Insects are the great opportunists in our fields and gardens, helping themselves to the food we planted. What they lack in size is more than made up for in diversity and sheer numbers. More creatures fall into the taxonomic class 'Insecta' than any other category. About a half of named living species on earth are insects, and still millions remain undiscovered, or at least without a scientific name. It is true that only a small minority could be classed as pests of agricultural crops, but they are a formidable foe.

Then there are the diseases. Fungi play an essential role in organic matter decomposition and nutrient cycling, but some like rust and mildew can be devastating for crops, particularly during wet weather. Next down in size are the microscopic bacteria, which often attack plants first damaged by insects or hail. Even smaller are the viruses. They are little more than DNA, and need vectors, usually sucking insects, to pick them up from one plant and inject them into another.

Plants can't run away from their attackers, but they are not defenceless either. Some go in for mechanical defences like thorns. Most are into chemical warfare and produce toxins to deter bugs and browsers from eating them. A potato hidden below the surface contains some toxins, but if part of the tuber is above ground it turns green. The green potato contains a particularly nasty toxin – the potato takes extra precautions against attack when it is exposed.

Plants can also defend themselves from disease attack. When the rust fungus invades a wheat leaf, a resistant cultivar can recognise the fungus and quickly kills its own cells surrounding the infection site to starve out the invader. This mini-scorched earth policy stops the rust from taking hold and limits the damage to the plant.

Whereas leaves and storage organs need to be protected from pest attack, most fruit are free of toxins when ripe. The colourful and tasty ripe fruit is an advertisement to a bird or animal to carry it off and thereby disperse the seed it contains. Sometimes the seed within the fruit is toxic or impenetrably hard, and so is discarded to germinate some distance from the mother plant. Other seeds rely on being eaten, carried away and deposited with manure. Their tough seed coats can withstand the digestive acids and enzymes in the gut. Some plants, like tomatoes and passionfruit, surround their seeds with a gelatinous pulp to help them slip past grinding teeth.

We get rid of some toxins during food preparation. A potato is inedible to many insects while under the ground, but the toxins it contains are inactivated by the heat of cooking. Next time when peeling onions with eyes burning, spare a thought for a bug chewing into an onion, whose antennae are just a few millimetres from its mouth. With cooking, onions reveal a rich sweet flavour.

Sometimes plant defences work in our favour. The pungent flavours of garlic and mustard are chemical defences. If you bake a whole clove of garlic, or eat a whole mustard seed, they have a rather mild flavour. But if you crush garlic or mustard you get an altogether more powerful effect. In both cases the plant stores a non-toxic precursor chemical next to a specific enzyme. When the grub chews into the cells, the precursor chemical and enzyme get mixed together, producing a powerful deterrent to further consumption.

The creatures we call pests have evolved ways of getting round the plant defences. We have helped them somewhat by domesticating food plants. For example, we select vegetables without the bitter or astringent tastes that make up part of their chemical defences. We have also selected for larger fruit; the modern large unsprayed apple is often infested by grubs while the neighbouring small crab apple is overlooked. Agriculture also simplifies ecosystems; uniform landscapes lack the diversity of the predator–prey relationships that can limit outbreaks of one particular species or another.

Domestication means we enter a new relationship with our plants and animals. We have selected plant traits that are desirable for us, but which reduce the plants' ability to survive or defend themselves. So we have to look after the domesticated plant, and this is what agriculture is all about. We plant at the right time, remove weedy competitors, and deal with the onslaught of insects, fungi, bacteria and viruses. If we do not deal with these freeloaders, they will eat deeply into our harvest.

Arsenic, the preferred poison for knocking off political rivals in the Middle Ages, was one of the first chemicals to use against agricultural pests. Apart from its toxicity to people, arsenic accumulated with repeated use, so it could never be a long-term solution. What was needed was something lethal to insects, but not so dangerous to people. The first synthesised chemical to fit the bill was DDT, but it turned out to be a disaster. Although it was far more toxic to caterpillars and mosquitoes than it was to mammals, it did not break down in the environment, and played havoc with the food chain, particularly among birds.

The search was on for chemicals that might be lethal for a short time, but would break down quickly and naturally into harmless by-products by the time we wanted to eat the food. The organophosphate group of chemicals, which kill by disrupting some biochemical pathways in the nervous system, mostly fit this category. The problem is that we, and many non-target species like bees and beneficial insects, share some of the same biochemical pathways. Such chemicals still pose a danger for farm workers because of their acute toxicity.

Some chemicals do seem to have genuinely low toxicity to mammals and short lives in the environment. An example is pyrethrum, a natural extract from a certain type of daisy.

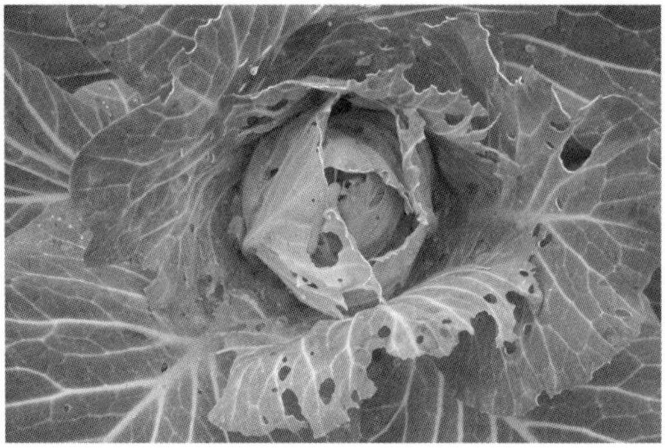

Caterpillars are eating machines and some have become resistant to the nasty chemicals that are used to eradicate them. Yet they can be easily controlled using preparations made from bacteria – sprays which have no toxic side effects on us or the wider environment.

Pyrethrum is a useful chemical, but it only kills a few important pests, and it kills non-target insects as well, some of which may be predators of the insect we are trying to kill in the first place.

The toxicity of agricultural chemicals is a hot topic. Each chemical has a toxicity rating, usually expressed as the amount it takes to kill a kilogram of rats. The test is called an LD_{50} – or the Lethal Dose to kill 50% of a population of rats. It is done this way because there may be a few particularly sensitive or resistant rats in a given population, which would skew the answer if we wanted the find the dose that killed all or none of the rats.

By law there is a maximum amount of chemical that can be applied to a crop and a 'withholding period' between spraying and picking. Each chemical has a 'maximum residue limit' – the maximum amount considered safe to consume – which has a high safety margin factored in. If these laws are adhered to, residues on the food we eat will be very low.

Of course there are problems. Unlike the LD_{50}, most of us are never exposed to a high dose of one chemical at one time. More likely we are exposed to a whole cocktail of different chemicals at very low doses for a very long time. Also, some people seem to be extremely sensitive to certain chemicals. Farm workers may be exposed to high doses, particularly those in poorer countries applying pesticides by hand without protective clothing.

The ultimate weapon against an insect is to find something that is totally specific to the target pest and virtually harmless to everything else. The so-called biological pesticides, of which *Bacillus thuringiensis*, or Bt for short, is the most well known. Bt is a natural soil dwelling bacterium of which there are several strains. One strain makes caterpillars – and only caterpillars – stop feeding when infected, and they soon die. The bacteria multiply in the dead caterpillar bodies and spread to others. Bt is harmless to all other creatures, including us.

Organic gardening books will give all kinds of homemade remedies to keep pests and diseases on the hop. I have tried many of these: concoctions of crushed garlic, chilli or tomato leaves to ward off chewing insects, detergent sprays for aphids, sour milk for mildew, glues for mites, sticky yellow paper for whitefly, and so on. Sometimes they seem to work, but in my experience there is mostly the therapeutic benefit to the gardener of at least 'doing something'.

Some say that if we look after our soils, then healthy soils will produce healthy crops, and insects do not attack healthy plants. The logic concludes that pests are a sign that something is wrong. I have often wondered on what basis a caterpillar would forego eating a cabbage because the cabbage was considered healthy. But there may be broad truth in the generalisation. Plants are continually synthesising sugar and amino acids which are used to construct new leaves, stems or fruit. When the plant is growing fast, the sugar and amino acid reserves are drawn down quickly, and there is not much up for grabs for the pests. Sickly plants tend to build up a bit of sugar, because it is not used up by new growth. This provides a super-meal for the bugs.

Recently I cut down a tough old camellia bush which had never supported a single pest. A few weeks later the cut stump sprouted and the new growth was covered in black aphids. Clearly the bush was now in an unbalanced state, with all these 'goodies' coming from the roots faster than they could be used. This evidently made the plant attractive to pests. I think I've seen sufficient evidence from my garden that luxury-fed plants are more vulnerable to attack.

Some pests in our garden, like the cabbage white butterfly, would wipe us out if they could, but we keep a step ahead of them with preparations of the wonderful Bt. But there are a few pests that cause 95% crop loss, like fruit flies in the late-maturing stone fruit and codling moths in the apple family. We have tried traps and lures and various cultural methods, but without sprays we are powerless to stop almost every fruit from being attacked. It is frightening what some pests can do. What if something as devastating as the codling moth arrived to attack tomatoes, corn, pumpkins or beans?

Commercial growers manage to produce unsprayed apples in our area. They do this by releasing pheromones – chemicals which the female moth releases to help the males locate her. The males can pick up minute quantities of the pheromone, and by tracking the concentration gradient locate the source. The farmers release pheromones all over the orchard, so the moth tracking system is overwhelmed and confused. Codling moths do not fly very far, so if you can disrupt mating then you can get their numbers right down. Unfortunately the method does not work in the suburbs – where every second backyard has a huge unsprayed tree, acting as a reservoir for new infection. I have now resorted to a netted enclosure for my apples and peaches, with a fine enough mesh to exclude the fruit fly and codling moth.

Our experience is that if we 'leave it to the balance of nature' we will lose a substantial portion of our crops, because despite our fantasies, agriculture is not strictly 'natural'. However, it is not wise to attack our foe head on. Take the aphid for example. You can spray them

Pheromone traps are a sneaky way to outsmart some pests. I use this fruit fly trap to find out when the pests are around. Then the fly proof netting goes down to protect the ripening peaches.

and kill 99.9%, but you will not kill them all, and they can bounce back at remarkable speed. When conditions are good almost all aphids are born female, can feed immediately, and within a week can have babies of their own. A female aphid can produce three to six young a day for several weeks, without even needing to mate. If you start with 10 aphids in week one, you can have 100 by week two, 1000 by week three and 10 000 by week four.

As the aphids start to feel crowded, they walk off to infest the neighbouring plants. When really crowded, they sprout wings and fly off to exploit a whole new area. Not only can the aphids reproduce asexually and develop a set of wings if they need them, they can also mate and lay eggs. These eggs can last through the long cold winter and hatch in synchrony with the new growth of spring.

Aphids employ ants to carry them to new succulent growth and even to protect them from predators. That is why you often see those large brown 'honey ants' crawling over your fruit trees. The ants are repaid for their service by a sugary substance the aphids secrete from their bodies.

With this explosive growth potential and mobility, the aphid will not be wiped out by intermittent chemical warfare. What we need is an ally in the fight that is on the job 24 hours a day. Fortunately aphids do have some enemies, particularly wasps. You aren't likely to spot the tiny aphid-destroying wasp, but it is easy to see their handiwork. The wasp lays one tiny egg into the body of the aphid. The little larva grows inside the aphid until it has consumed it all, and then pupates. The young wasp emerges from the pupa and then cuts a small circular hole in the top of the aphid and flies away. If you look carefully on aphid-infested plants you are sure to see the brown mummified bodies of aphids with a hole in the top that were eaten from the inside by wasp larvae.

While wasps attack from the air, ladybugs roam the foliage like armoured vehicles, picking off the strays. The larvae of the lacewing lie in ambush, stuffing their fat bodies with the fat sweet aphids. Spray the aphids and we may just wipe out these allies, and the pest always bounces back faster than the predator. And each generation that comes forth, by virtue of their parents that survived the last spraying, have a little more resistance to the chemical.

The global use of insecticides and fungicides is increasing each year. At the same time the number of damaging pests that have become resistant to chemicals also goes up. Meanwhile the number of pesticides available to kill pests is going down, as some of the older chemicals are deregistered for use on health or environmental grounds. It is not a pretty picture.

Plant breeders are in the front line of battle against diseases, developing new varieties that can stand up to the evolving new virulent strains. This has been remarkably successful, particularly for the staple food crops. Breeding for resistance to pest attack has been more difficult. Controversy aside, the splicing of the Bt gene into cotton and then other crops has been incredibly effective against caterpillars.

Plant resistance must be combined with the appropriate farming practices. Wheat farmers in southern Australia found that they often got higher yields when planting wheat after mustard seed or canola. It turned out that diseases had been attacking the roots of the wheat crop. Long before we see the wheat coming to head, the yield potential has been trimmed by disease. It turned out that mustard and canola produced chemicals that acted as a biofumigant, clearing the soil of these diseases. Many farmers blamed the lack of rain for low yields, but in fact the problem was a weak root system not able to extract water from deep in the subsoil.

This comes as no surprise to vegetable gardeners. They know that the tomato family share the same diseases and should not be planted year after year in the same patch. Turning water into more food will always mean finding clever rotations to beat the soil-borne diseases.

It is hard to say how much of the world's food is destroyed by pests and disease, but it is very substantial. Even after the harvest is in, the crops are not safe from attack. It is not unusual for farmers in South-East Asia to lose a quarter of the crop to rats – often more. Someone once calculated that rats in Pakistani villages ate 300 million tonnes of rice each year, and damaged a lot more in the process.

We will never wipe out the pests and diseases and we don't need to. But we do need to know our enemy, its biology, lifecycle, predators and weak links. There is a threshold for any pest that causes acceptable damage, and with crop choice and cultural methods plus biological controls we may be able to outsmart most of them most of the time.

Chapter 7

Flavour

When we go on holiday, we buy a pack of assorted breakfast cereals comprising six different types, each in their own small box. It has become a tradition in our family. Youngest gets first pick, and invariably grabs the cereal box with the highest sugar content. All three kids are satisfied on day one because there are three cereal boxes containing over 36% sugar. By day two the stakes are higher, because the remaining three boxes in the pack range from 9 to 32% sugar, and that's worth fighting over.

Some of these so-called breakfast cereals have more in common with lollies than the rice, wheat or corn they are based on. The most basic of breakfast cereals, like Corn Flakes and toasted rice, are highly processed versions of maize and rice. Cereal grains are mostly made of starch, which is the energy source for the germinating seed. Starch on its own does not have much flavour, so to make Corn Flakes a bit more enticing, they have about 10% of sugar added, plus a dash of salt.

For a genuine cereal breakfast we need something like compressed biscuits of wheat, such as Weet-bix. Weet-bix is made up of 97% wheat, so has almost the same composition as the wheat seed: 12% protein, 64% starch 11% fibre and 1% fat. An extra 3% is added sugar, and the remaining 9% is mostly made up of water. Although Weet-bix are very good for you, they are not very palatable on their own. We need milk to turn these biscuits of wheat into a hearty breakfast, and milk contains 3.5% fat and 5% sugar (lactose).

Adults don't usually have quite the same sweet tooth as kids. I prefer muesli for breakfast, which is more satisfying than bland old Corn Flakes. Our toasted muesli has double the protein and half the sugar content of Corn Flakes. But the real difference between Corn Flakes and toasted muesli is the fat. Muesli, toasted in vegetable oil and containing nuts, sunflower and sesame seeds, can easily contain over 20% fat.

Toast is pretty similar in composition to Weet-bix, but is rather dull on its own. First we add butter (fat) and then jam (sugar) to turn it into a tasty breakfast. It is the same with the starch-filled potato. We fry them in oil, cover them in gravy or, in the case of a baked potato, add butter or sour cream. Although we can get all the fuel we need from the complex carbohydrates in the wheat or potato, our taste buds seem to drive us towards adding fat, sugar or both.

Some people, usually kids, smother all kinds of food with tomato sauce. Once again it's the sugar they're after; a mighty 28% plus 3% salt for the brand in our fridge. Plain boiled rice makes a pretty dull meal, but it can be transformed by soy sauce. There are flavours in the fermented soy, but salt is a big player; soy sauce is a 17% solution of sodium chloride. When we analyse fast foods we are usually being seduced by some combination of elevated fat, sugar and salt.

Strangely enough, our chickens show a similar zeal for sugar and fat. They have a self-feeder containing layer pellets with 16% protein and 6.5% fat, the perfect ration for laying eggs. They also have some grain thrown into the litter which they enjoy scratching for. Despite constant availability of high protein food, they all charge the fence when they see someone approaching with the kitchen scraps bucket. They will fight over bacon rinds or fatty offcuts of meat. When cake is thrown out, the chickens devour it with such gusto they must scarcely taste it.

The fat, sugar and salt trick has been so successful that it fills the supermarket shelf, fridge and freezer space. Processed food is so convenient, easy to store, quick to prepare and sometimes tastes good too. The trouble is that it's not the healthiest way to eat, for our bodies or for our environment. And most processed food has a huge energy footprint, particularly with regard to processing and packaging.

Since it is so good for us to get most of our energy from starchy cereals, what drives us to go for the sweeter, fattier options? One explanation is that sugar and fat give us quick and 'cheap' calories, compared to the labour involved in collecting and preparing a starch-based meal. Such quick calories would be hard to pass up for the early farmers and hunter–gatherers.

I recently tried to demonstrate this to our youngest, who has a fondness for very refined food. We had a plot of wheat growing in one corner of the garden. We cut the wheat with a sickle and collected up the stalks into bundles. Then we began the laborious job of bashing the heads to dislodge the grain. An hour or two later we had a bin of grain, chaff and bits of straw.

Despite all I read about exotic foods and the novel creations of celebrity chefs, nothing tastes better than freshly picked vegetables roasted in the oven with olive oil.

Next we winnowed the grain by tossing it into the air so the lighter chaff would blow away. No amount of winnowing gives totally pure grain; there were some seeds encased in their husks and heavier pieces of straw that needed to be removed individually by hand. Then we ground the grain in a small hand-operated stone mill, progressively loosening the setting between the grinding stones as exhaustion set in, because it was easier to mill a coarse rather than a finer flour.

We added oil, salt, yeast and water to the flour, kneaded the dough and had a small loaf ready for the oven. The energy we had expended thus far may well have exceeded what we would recover from eating the bread, and we did not even have to gather the wheat from the open fields or collect the wood for a fire.

Imagine if, as early farmers on the way to the fields to collect wheat, we had come across a bee hive in a tree, or managed to snare a small animal. The sweetness of honey would be a rare treat. We would roast the animal and our hunger would drive us to not waste a single piece of fat. But sweetness and fat would be relatively rare in our starch-based diet. Modern agriculture has changed all that. Sugar and fat are very accessible indeed, and we fall for them over and over again.

The way out of the sugar-fat-salt trap is to rediscover the myriad of flavours found in the garden. It is not that we need to stop eating sugar and fat; just that we need to get better all-round value from that which we do consume.

When we eat fruit, we get the sweetness we're after, as well as vitamins and antioxidants and fibre. But for a really good strawberry, we're after much more than sugar. Sometimes a strawberry tastes sweetish, but the flavour is bland. Sometimes the fruit is hard and the flavour a bit tart. And sometimes we savour the perfect strawberry.

The flavour of the strawberry is principally made up of three components. There is the acid which gives the tartness and the sugar which gives the sweetness. The third group of components are flavour aromas. One is called furaneol and smells a bit like caramel and the other is mesofurane which smells a bit like sherry. Mixed together, they make the authentic strawberry flavour, and with the right balance between the acid and the sugar give us the perfect strawberry.

The problem is that the temperatures needed to get the right amount of sugar are higher than the temperatures required for the right amount of flavour aromas. The perfect strawberry needs coolish nights, say below 15°C, to make flavour, and warmish days to make sugar. What this means is that the perfect tasting strawberry can only be grown in some places and at certain times of the year.

If we want a strawberry crop year round, then we need to target warmer areas, which are going to produce strawberries low on flavour. And if we need to transport them long distances to market we need to pick them a little unripe, when the acid is a bit high. Here is the great dilemma of the industrial food chain. The perfect looking year-round strawberry will usually be grown under very different conditions needed for the perfect eating strawberry.

Most people are well aware of this, perhaps because we still have a generation who ate substantial amounts of fruit and vegetables from someone's backyard garden. 'Home-grown food tastes much better' is a near-universal cry.

This is a bit of an over-simplification. I can't taste the difference between our home-grown onions and potatoes and those from the shop. Most of the time I can't tell the difference in the citrus either; ours is sometimes worse flavoured than shop-bought. But there are some fruit that are not even comparable. Home-grown apricots are so delectable they may as well be a different fruit altogether from the ones we buy.

The difference between shop-bought and home-grown comes back to how the fruit ripen. Oranges and lemons, for example, both contain the same amount of sugar. Oranges slowly lose acid as they ripen, so the sweeter taste comes to the fore. The slow change means that there is a long period over which an orange is tasty and the actual time of harvest is not critical. Lemons don't lose their acid as they ripen, and so retain the sour taste. Both reach our homes tasting as good as they should.

Apricots do not become sweet by losing acid, but by packing sugar into the fruit in the last couple of weeks before harvest. If you pick an apricot early, it tastes bland because it has not had time to accumulate all its sugar. On some of our trees, the side of the apricot facing the sun is slightly riper than the other side, so there is never an optimum time to pick. Overall, the later you can pick the better, and that does not work for the food chain which needs time to pick, sort, pack, transport and market the fruit before it starts to soften. For the best tasting apricot you need to be lying under the tree with your mouth open.

Bananas ripen not by gaining sugar or losing acid but by converting starch reserves into sugar. This is very useful, because the fruit can be picked when totally green, full of starch but with no flavour at all. Then, during the processing chain the fruit is exposed to a chemical called ethylene, and kept at just the right temperature. This stimulates the breakdown of starch into sugar.

Quite a few fruit are chemically ripened in this way. Ethylene is a natural chemical that the plant itself produces to stimulate ripening. The food processing industry has learnt how to mimic the plant strategy exactly, so fruit can be picked and transported while still firm and unripe, and then brought to full flavour at the right time.

Apples, avocado and kiwifruit are some of the fruit ripened in this controlled way, and so they generally reach our homes with good flavour. Peaches, plums, apricots and all the berries, which are best ripened on the tree or vine, will almost always be better home-grown.

There are other reasons why the home-grown product is tastier. Commercially grown varieties are often bred for their ability to withstand the rigours of packaging and transport, not for their flavour. Going for maximum yield can also come at the expense of flavour. I once carried out large-scale experiments on the effects of water stress on tomatoes. Withholding some water during ripening dropped the yield slightly but improved the flavour immensely.

Delicious as it is, fruit does not satisfy genuine hunger. No restaurant has a fruit-based dish in the list of main courses. Almost every main dish is based around meat or seafood. As vegetarianism becomes more mainstream, there will always be a meat-free dish or two on offer, but rarely more than 10% of the menu. When customers want the best meal possible, it seems that meat is hard to beat.

At university, we invited international students to a Christmas meal, and they came for the roasted goat, not the roasted vegetables. The goat contained something like 20% protein and 30% fat. Such protein- and calorie-rich foods are hard to produce from the garden. We lived on a small holding at the time, and the few goats inhabited an area many times bigger than the large vegetable garden, and we only killed one a year.

My present garden does not produce a large proportion of the 2500 Calories and 70 grams of protein I need each day. We haul a large bulk of vegetables into the kitchen, but much of it is water. A salad of lettuce, tomato and cucumber would contain 95% water and less calories than the salad dressing. Vegetables like zucchini seem to grow while you are watching them

The wood piled up under this mango tree is from other mango trees. They have been felled to make space for more maize and to provide fuel for cooking. Even the most delightful of fruits has to make way for the bland staple food in a time of crisis in this sub-Saharan country.

– too small to pick today and then too big tomorrow. In fact they are filling up with water, not food calories. Watermelons grow more slowly because they are accumulating sugar, and pumpkins grow slower still as they are accumulating sugar and starch. The more calories something has in it, the longer it takes to produce.

Take cheese for example. Our dairy goat consumed a vast amount of grass. We collected a few day's worth of milk then added rennet to curdle the milk and separate the whey. We added salt and then used a homemade press to remove the remaining moisture. A week later we had a small block of mildly flavoured cheese. Cheese is a concentrated food source; grass is concentrated into milk by the goat, the protein and fat are concentrated in the kitchen by separating the whey and the remaining water forced out under pressure.

The concentrated food from animals requires more land and is therefore more expensive. Most poor people therefore rely on starch-based diets, usually from a cereal or sometimes a starchy root crop like cassava or potato. The bland flavour of starch foods needs to be overcome and this is usually done through some kind of sauce. The sauce would contain some oil and then vegetables like onions, tomatoes, eggplant, carrots or cabbage, which add bulk and subtle flavours. The sauce then also carries the stronger flavours of herbs and spices, chilli and garlic.

Such meals may have a small component of meat, or more likely beans, lentils or chickpeas. It seems there may be health benefits from such diets. Our bodies can digest almost all the meat we eat, leaving little to pass through the system. If there are any carcinogens in the meat, perhaps formed during cooking, they linger in the gut. Vegetables are mainly composed of cellulose, which forms the cell walls of plants. We call this roughage or fibre, because we can't digest it. Roughage moves through our systems quickly, sweeping the gut clean of any nasties.

The summer garden presents a permanent snack on demand. I would never walk into a shop and buy a tomato, but the garden tomato is delicious, and there are usually three or four varieties to choose from. Similarly a shop plum is a rather dull affair, but in the garden the plum is a delight. We grow four varieties of plums and they get tastier as the summer wears on. The first of the season is the Japanese variety 'Santa Rosa', with red skin and soft, juicy, yellow fruit. Next comes the blood plum 'Satsuma', with its firmer, more sharply flavoured flesh, then 'Angelina', a European plum with a deep purple skin and firm, very sweet yellow flesh. Last is the greengage which looks least appetising but tastes superb.

Eating from the garden is a different experience from eating what you buy. You can select fruit of slightly different maturities, depending on whether you are after a tarter or sweeter flavour. A bunch of grapes in summer is nice, but I like to mix up three or four varieties with different flavours. We grow the sweet 'Flame Seedless', the muscat-flavoured 'Italia' and 'Isabella', a labrusca-flavoured black grape with a sharp contrast between the skin and flesh. Really good fruit keeps the craving for junk food at bay.

The journey to a healthier diet will have to be via the flavour route, and we are having some success. At home we try to build meals around what is plentiful in the garden, using meat more sparingly as flavouring. Everyone loves roasted vegetables, particularly Desirée potatoes, butternut pumpkin and certain types of red capsicum. Bruschetta is a favourite; tomatoes blended with red onions and basil on toasted French bread basted with oil. Similarly everyone loves home-made pizzas with loads of onion and home-made tomato paste loaded up with all manner of vegetables and smothered in cheese.

In winter we get through huge amounts of Chinese greens, Tuscan kale, broccoli and cauliflower, usually mixed up in stir-fries with sweet chilli sauce. It is not quite so easy to get through as much cabbage, but we manage it fried in olive oil and seasoned with soy sauce. Yes, we still need the sugar, salt and fat, but we have got a lot more home-grown roughage into the family diet.

I read recently that we had to rediscover simple traditional cooking to save both our own health and that of the environment. Traditional cooking from places like the Middle East and parts of Asia developed around scant access to meat and fat, so their local cuisine developed amazing ways to make such simple foods palatable and full of flavour. To make really flavoursome food from simple ingredients requires time and skill, and these are the two ingredients hardest to find in a busy world.

PART 2
A JOURNEY THROUGH SOIL

Anyone who has studied soil science at university will tell you that understanding soil and water is difficult. Sandy soil is such different stuff from a clay soil that it's amazing we call them both 'soil'. Keeping track of water requires expertise in measurement and mathematics, but understanding the soil system as a whole requires chemistry and biology too. Irrigation can take a heavy toll on river ecosystems, so the soil scientist needs to be an ecologist as well.

I studied soil science, but my real learning only started with my first job when I was thrown in the deep end. I needed to know how much water was in the soil, and I discovered it was not easy to be a good irrigator. Then I worked in Africa and got involved in developing new devices that could perhaps do the job more simply.

This collection of essays is my own journey of learning about soil and water, and how to make irrigation more sustainable. In the early days of my research career I thought I could revolutionise plant production with some rather 'way-out' ideas. But I discovered that making a difference takes more than just good ideas. People are also part of the system and to make a real difference you need to take people on the journey with you.

Soil fascinated me from the first day I started growing vegetables. The soil stores water and nutrients, a bank from which the plants make their daily withdrawal. Soil can be improved, but it can also be degraded by salt, waterlogging and compaction. Most soils can produce fantastic crops with tillage, fertilisers and irrigation water, yet the farming practices which are successful in the short term can play out differently in the longer term.

Clever Clover was my first attempt to translate ideas from the garden to the commercial world. Subterranean clover grows during the winter months, and it dies down naturally in summer to form a mulch on the soil surface. Being a legume, clover added nitrogen to the soil, and as the roots decayed they left interconnected pores through the undisturbed soil. On paper it was the perfect no-till, no-weed, no-fertiliser system, but the real world held other challenges. A complementary system using lucerne proved more robust.

It is not possible to be a good irrigator without understanding how much water is held by the soil. A whole lot of devices for measuring water in the soil have been invented over the last two decades and I've used most of them. Some irrigators rely on these tools, but most don't, and you almost never see a soil water monitoring device in a garden.

The Wetting Front Detector was designed to be a soil water monitoring tool that worked in a way people could visualise. The funnel was buried in the root zone. If sufficient irrigation water percolated past, the funnel collected some water which activated a float, visible at the soil surface. The water sample was also used for measuring salt and fertiliser levels in the soil. I cut the shape of the detector out on sponges to demonstrate how wetting fronts move, and how the detector collects water.

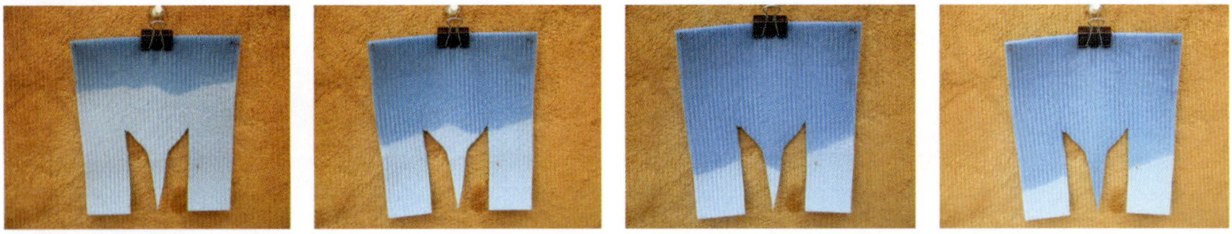

Chapter 8

The tale of Clever Clover

After finishing my agricultural science degree, my first job was to grow vegetables. It was hard to believe I would be paid full-time to do my hobby. A year or so after I started my job at the university research station, I stumbled across an idea that I wanted to throw all my energy into. I started a PhD, set up experiments, collected the data, and bit by bit pieced together what I thought was a very interesting story.

Indeed, I thought I was onto a vegetable growing breakthrough. I submitted a short summary of my work to a conference, but was not invited to give a paper. Few scientists visited our research farm, outside our immediate research group. I wanted to tell the whole world about the work I was doing, but not many people seemed interested.

Several years later, after my postgraduate studies had been completed and I was working at CSIRO, I was interviewed by a science journalist. He wrote a single page media release on my PhD work. The press release suggested my approach would lead to a revolution in vegetable growing and came with the eye-catching title of 'Clever Clover'. Having mellowed substantially since the project started, I suggested the media release overstated the story a bit. 'Trust me', he said, 'this media release will drop onto the tables of sub-editors where it gets a 10 second glance before being filed in the bin. You have to get their attention'.

The lettuces on the left are growing on a compacted soil that has not been cultivated. The ones on the right are clearly enjoying the softer soil that allows better root growth, giving them access to more water and nutrients.

The story ran in leading newspapers of most Australian capital cities and at least 35 other papers around the country. There were seven TV appearances of one sort or another and I lost count of the radio interviews. My face was on the front page of two magazines. The postman struggled up the stairs at my new workplace with his bag bulging. There was not space in all the pigeon holes in the building to put the mail that day. He just tipped the mail in a huge pile on the floor, and left muttering that the strap on his mail bag had broken. Letters continued to arrive by the hundred.

So, what was this magical system called 'Clever Clover'? The story began with my job at the university, which was to test the idea of irrigating vegetables using only solar power and drip irrigation. It was the mid-'80s, drip irrigation was still a new technique and photovoltaic solar cells were rarely seen out of a research setting. We grew vegetables on raised beds, which the tractor straddled with its wheels in the furrows on either side. In the centre of the bed was the drip tape, buried about 10 cm below the surface. Every 30 cm there was a small emitter, delivering 300 ml of water per hour.

Drip irrigation looked like the answer to a whole lot of problems. When we sprinkler-irrigated our young lettuce seedlings, the whole field got soaked, including the furrows where nothing was planted. Most of the water evaporated from the soil surface within a few days. During warm and muggy weather, fungal diseases spread rapidly through the wet canopies. When the rain came, water ran off the beds and flooded the compacted furrows and the tractor wheels churned up the mud as we tried to spray the crops. Weeds grew everywhere.

By contrast, the drip tape supplied water only to the centre of the bed, and lettuce roots grew towards water. Ideally we had a wet cylinder of soil around the buried drip line, meeting the soil surface in a narrow strip and wide enough to satisfy the plants growing in rows on either side. We did not lose huge amounts of water directly from the soil surface and we never wet the foliage.

Weeds found it harder to establish, because the edges of the beds and furrows remained dry. These dry areas soaked up the rain when it came, so the fields did not get so boggy in wet weather. Fertilisers were dissolved in the irrigation water and delivered directly to the plant roots.

Running the whole system by solar power completed this ideal scenario. Plants require less water under drip irrigation than sprinkler or flood because there is less evaporation from the soil and because drip irrigation waters evenly with no wet or dry spots. Drip irrigation also runs under much lower pressure than sprinklers. Pumping less water at lower pressures means less power.

Solar power and drip irrigation were a perfect match. When the sun's rays hit the solar panels in the early morning, the photovoltaic cells started to produce electricity and turn the pumps. The plants too started to transpire. The sun's energy in driving transpiration and in producing electricity was pretty much the same; we just had to match the area of panels to the area of the field. It is true that plant water use also depends on the temperature, humidity and wind speed, but the basic relationship held. The long, hot days extracted more water from the plants but gave us more power to pump, and therefore replenish the soil water reserves. We could not pump much under heavy cloud, but we did not need to, as it was likely to rain.

I was the research assistant in charge of the day-to-day management of the project and this job encompassed three of my passions: understanding soil, understanding water, and growing vegetables. There was only one catch. We were testing the performance of different solar pumps and for that we needed to know what area a pump/panel combination could irrigate. This meant that the fields were much bigger than we normally use for experiments. But after so many years of growing vegetables in the backyard I felt up to the challenge.

The reality of the situation only hit home when I put in my first order for seedlings. I had bought punnets of seedlings countless times from nurseries and teased apart the eight small plants. There are few more satisfying tasks than settling young seedlings into freshly-dug soil and watering them in. My first order in my new job was for 24 000 lettuce seedlings. This was no place for garden forks and watering cans. The land was prepared by chisel plough, disk harrows, rotary hoe and the seedlings put into the ground using a mechanical transplanter, all of which I was using for the first time.A week later the new seedlings had established, but the beds showed a tinge of green. Millions of weeds were germinating. I got out my hoe and carefully stirred up the soil around the young plants. An hour later at the end of the first row, blisters were forming on my hands. Only another 29 rows to go. This was my rude introduction to the scale of agriculture that the self-sufficiency books I read as a teenager had not mentioned. No commercial lettuce farmer out there will be impressed by a crop of 24 000 plants. Even planting that many seedlings a week would not make you a particularly big grower.

Nine weeks of irrigating, fertilising, weeding and spraying and the crop was ready for harvest. Picking was a lot harder than planting. Lettuce are 95% water and water is heavy. A local vegetable-growing Maltese family helped out. We had to start at first light while the plants were still covered with dew. The Maltese children sped up and down the rows, bent double, cutting and packing 12 lettuce to a waxed cardboard box, which were piled high on a truck. I was relatively fit, but could not keep up their pace.

I had produced my first commercial crop, and the Maltese farmer said it was a good one, but bigger challenges lay ahead. We had to plant a tomato crop on an area double the size, and it was still under grass. I started with a mouldboard plough, which sliced under the roots and turned great sods onto their side. Then I dragged the iron teeth of the chisel plough into the subsoil to break up the layers of compaction.

Everyone knows what a good soil looks like, but it is surprisingly difficult to describe in scientific terms. When we pick up a lump of soil and it breaks down easily in our hands into soil 'crumbs', we talk of a friable soil. If a lump of soil can't easily be broken down into 'crumbs', we call it a clod, and it is nearly impossible to germinate a seed in cloddy soil. If a lump disintegrates into its constituent clay, silt and sand particles it looks like dust when dry and mud when wet.

The time to work the soil is when it is moist because then the lumps are most likely to break along their natural planes of weakness into soil crumbs. Plough the soil when it's too wet and it smears like putty and dries like a brick. Plough the soil when it's too dry and it shatters into fragments which disintegrate when wet. In my case I ploughed the tomato lands too dry. The chisel plough dragged up large clods which had to be broken down by force, back and forth with the disc harrows and rotary hoe until we could form beds of soil fine enough to transplant into.

Eventually we had 50 dead straight rows, 100 m long with 80 cm wide beds between two wheel tracks and a line of drip tape buried in the centre. We were about to plant out the thousands of tomato seedlings when it started to rain. It rained for eight days with no let up, 192 mm and the wettest spring on record. The soil turned into a red soupy mud. The beds slumped into the furrows. Each day after the clouds cleared we ventured onto the fields to see if we could plant, but our boots sank into the sludge and we slipped and slithered our way back to the safety of the headland. Trapped for days in little seedling trays, the tomato seedlings lost their lush green appearance. First they elongated, then they turned a sick looking bluey-grey.

I took a few days off work and returned to my vegetable garden. At that time I rented a small two-roomed wooden shack on a dairy farm near the research farm. For the first time in my life I had a garden far bigger than I needed. Until now my farming instincts had been somewhat curtailed by lawns and shady trees in the suburbs, but the friendly farmer had fenced off a huge area right next to where the manure-laden water flowed out of the dairy.

We showed that many vegetable crops could be grown no-till, as long as we used mulches. In this case the decaying clover roots left some pathways through the soil for the lettuce roots to follow, and the mulch kept the soil cooler and wetter.

Some time earlier I had sown a cover crop of clover in the part of the garden I did not need, using some seed I found split from a bag in the dairyman's shed. The clover grew thick and fast, covering the ground in a dense sward and smothering weeds. I had planted subterranean clover, 'sub clover' for short, a plant native to regions with cool wet winters and long dry summers. The clover survives by burying its seed at the end of the rainy season and then germinating when the rains return after the dry summer.

Sub clover grows in regions where the winter rainy season ranges from four to six months. Different varieties have adapted to complete their lifecycle before the summer dry kicks in. They do this by timing flowering to a specific time of the year, using day length as a cue. It takes about six weeks to go from the day the flowers appear to a viable seed, so it makes sense to flower when there are likely to be at least six weeks of moisture left in the soil. Once the seed is mature the plant dies down. This makes sub clover a rather unusual plant, in that you can predict – more or less – the time of year each variety will die.

In my case the clover had turned brown and collapsed to form a thick mat on the soil surface. I lifted this mat of dying vegetation and scooped up a handful of soil below. Despite not being ploughed, the earth was soft and sweet-smelling and the soil crumbs fell through my fingers. Unlike the sludgy soupy mess at the experimental farm down the road, my garden soil was riddled with earthworm holes and little channels made by decaying roots. Presumably the excess water had drained into the subsoil. The soil looked in perfect condition for transplanting tomato seedlings, unlike the soil I seemed to have destroyed on the experimental farm.

This was the start of the idea that would become Clever Clover, an idea which for a brief moment became a media sensation. Clover is a legume, taking nitrogen from the air, turning it into a form plants can use and then returning this vital nutrient to the soil when the plants decompose. The clover roots grow and die, leaving interconnected pores that permeate the soil. The organic matter feeds the soil organisms. The leaves form a mulch which protects the surface soil from being smashed by raindrops and baked by the sun. Under the mulch, the soil temperature is cooler during the day and a little warmer at night. The mulch is a barrier to evaporation and best of all, a barrier to the weeds. We could dispense with the plough, the nitrogen fertiliser and the herbicides, all of which seemed pretty clever to me.

In any event, the title 'Clever Clover' was a deliberate play on the government slogan at the time of 'Clever Country', which aimed to promote investment in science. The media release had a picture of a gardener slouched in a deck chair while the tomatoes just grew in front of him. With no need to dig, fertilise or weed, there was nothing else for this Clever Clover gardener to do, other than relax. I used the Clever Clover system in my backyard garden and it made wonderful footage for popular garden TV shows.

One of the TV programs wanted us to produce a pamphlet describing how gardeners could set up their own Clever Clover garden. It would not be easy for gardeners to get the right varieties of seed, so we decided to make a Clever Clover kit. It contained seeds and rhizobium, the special bacterial inoculants that help legumes to get nitrogen from the air, and instructions on how to set up a Clever Clover garden.

Clever Clover sold like hot cakes. We sold 15 000 of the $10 kits before we stopped counting and then passed the idea on to a third party a year or so later. Fifteen years on they are still available for sale; someone somewhere is still setting up Clever Clover gardens.

But Clever Clover never had the influence I once hoped for in the real world of commercial vegetable production. The problem was that any new innovation developed to solve one set of problems throws up its own new set of problems. What if the clover cover crop gets infested with weeds or slugs? How do you plant through the mulch? How to manage crop residues after harvest in a no-tillage system? There were a host of other questions. The idea made sense, but it was hard to implement on a commercial scale.

Clever Clover was my first attempt at intervening in the way we produce our food and it left me with two enduring lessons. First, there are powerful forces shaping the way food is produced, and I deal with some of these in later chapters. Everything about Clever Clover sounds right, and in the long term we will have to find systems that better protect the soil, build up organic matter and have less reliance on chemicals. But the recent history of agriculture reveals farms getting bigger, specialising on single or only a few crops, and increasingly reliant on machinery and chemicals. Before we can intervene successfully we need to understand these forces. Ecological approaches such as Clever Clover require enormous skill, are harder to manage on a large scale and in the short term carry more risk. Farmers have enough risk to contend with already.

Second, there is a big difference between having an impact and having an influence. Clever Clover made a big impact. We did not plan it that way, but it did make a perfect story for the mass media. You could not miss the headlines. But Clever Clover had much less influence. To be influential, an idea must take root in the minds of people and change the way they think about things, and eventually the way they do things. The ideas at the heart of Clever Clover need to be tested and adapted and improved for local situations over many years. Clever Clover was delivered as a recipe for the no-work, eco-friendly productive food garden. In reality, that recipe does not exist.

The ultimate lesson was that, if you really want to change something – to have enduring influence – you need to stick at it for a long time. This is difficult in an environment when the investment cycle is short. The prevailing culture is skewed towards short-termism, going for the immediate impact and convincing ourselves that just because something makes a splash, it will deliver in the long term. For the next invention I worked on, I made up my mind I would persevere long past the time the initial enthusiasm and success had worn off.

Chapter 9

Soil and soil scientists

I have always loved watering the garden. I love the lushness after rain, when the air is warm and humid and the plants are luxuriant with sappy new growth. Then, as the dry weather takes hold, the leaves droop under the sun's intensity. A wilting plant is losing more water from its leaves than it can suck out of the drying earth.

In an attempt to lose as little water as possible, the wilting leaf closes the pores in its leaves tightly. This helps to keep the water in, but it shuts out the carbon dioxide from the surrounding air, and so starves the leaf of the raw material from which to synthesise the sugars for new growth. Still, some water escapes the waxy layer covering the leaf surface, and the once-turgid leaves hang limply. The plant has effectively shut down, trying to avoid complete desiccation until the next shower of rain. Turn the tap on, and the recovery is almost instant.

I always considered myself to be quite a good irrigator. I watered my garden once a week or so in summer, and when I saw signs of wilting I watered more often. It was only after I started measuring water in the soil as part of my first job, and later as part of my postgraduate studies, that I found out how difficult it was to be a good irrigator.

Seventy per cent of all water used in the world goes to watering plants and growing food and fibre, and most of us who use this water have little idea how complex an activity we are mixed up in. No doubt large amounts of water are wasted by irrigators, but in this resides a great opportunity, for we can become a community that understands water much better. With a bit more know-how, it is possible to produce more food with a lot less water.

The subjects in the first year of an agricultural science degree follow on from what we were used to at school: maths, physics, chemistry and biology. Then you move on to more specialised areas like genetics, physiology, microbiology and the study of insects and disease. Soil science is one of these specialities, and although it is a mixture of the core sciences we learnt at school, it is the one subject that trips up more students than any other.

Soil science can be tackled from three different directions. Soil physics describes the movement of water and air and the growth of roots through soil. Soil chemistry is about nutrients and toxins, and soil biology is about the microbial life in the soil. The three branches of soil science are intertwined. We can't say for certain that a certain soil is good or bad because a whole range of soil properties combine to determine how a plant will grow under a range of conditions.

Soil scientists tend to fall into several groups. There are the mathematical types, who see soil as a porous medium governed by complicated equations. They spend a lot of their time working on computer models. Then there are the muddy boots types who like to measure, and who all too frequently find their measurements do not tally with the mathematical predictions. Lastly there are the lay soil scientists, with innumerable theories on how to manage a particular soil, and who are hard to prove wrong because no two soils behave exactly the same way.

Fortunately there are some basic truths that unite us all, and which are fundamental to becoming a better irrigator. Here follows a simple soil physics lesson that can be conducted while having a bath. Drop a facecloth in the water and then hold it up by two corners so it hangs as a square. Water will trickle from the base of the cloth, fast at first and then drop by drop. After a while the drops get further apart and, if you wait long enough, will eventually stop altogether.

Then squeeze the cloth gently and 20 ml of water are expelled. Squeeze harder and we get another 20 ml. Squeeze tight and a further 20 ml are pushed out. After this we can squeeze till our knuckles are white, but we can't get any more water out.

Soil holds water in much the same way that the facecloth does. The cloth is saturated when first lifted up from the water. When we suspend the cloth from our fingers, gravity drains much of the water out, but not all of it. The remaining water clings to the cloth fibres or gets held up in small pores between the fibres. In other words the force between the water molecules and the cloth fibres is stronger than the downward pull of gravity.

A gentle squeeze can dislodge some of this water held by the cloth. Squeezing is analogous to a plant sucking water from the soil. Plants have to exert a force – a sucking force greater than the force of attraction between water molecule and cloth fibre – to obtain their water.

The important point is that we have to squeeze harder and harder to get each successive 20 ml out of the cloth. We can visualise this by placing the wet cloth under bricks. Half a brick squeezes out the first 20 ml, two bricks expel the next 20 ml and eight bricks are needed to get the last 20 ml out. This is one reason why irrigators err on the side of putting on too much water. Once the soil starts to dry, it is very hard for the plants to get the water out.

No two soils are the same, largely because each soil is composed of different sized particles. The biggest size we call sand, the next size down is called silt and the smallest particles are called clay. The size of the particles determines the surface area as well as the average size of the pores in a given soil. Large particles pack with large pores between them and small particles pack leaving a network of tiny pores. Water molecules are attracted to surfaces and fill up smaller pores, so particle size has a huge influence on how much water a soil can hold.

If we start with a piece of rock shaped as a cube, 1 cm wide, long and high, we have six sides each of one square centimetre giving a surface area of 6 cm^2. Now if we divide this cube into 1000 smaller cubes each with 1 mm sides, the surface area goes up to 60 cm^2. These 1000 cubes would be about the size of coarse beach sand, which is not really soil.

If we take our cube and divide it up into tiny pieces with one thousandth of 1 mm sides, we would have a one thousand billion particles (1 million x 1 million). These thousand billion particles would be roughly the size of clay particles, and have a surface area a thousand times greater than the sand. In fact, clay particles are shaped more like tiny flat plates than tiny cubes, and so have a greater surface area than if they were tiny cubes. One type of clay has a surface area bigger than five football fields for every teaspoon full. Since water and nutrients cling to surfaces, this has a big impact on how a soil behaves.

If the cube of rock we started with was quartz, it would weigh about 2.6 grams, giving a density of 2.6 grams per cubic centimetre. Most soil has a density just over half of rock, because soil particles pack with air gaps between them. The organic matter and other substances in the soil also stick particles together in ways that give it an open pore structure. Structure is a vital component of soil, because it is the pores that contain water and air, and where the roots grow. A good soil is only about 50% solid.

When we lifted the facecloth up out of the bath water, it was saturated. All the spaces between the fibres were filled with water. When the suspended cloth had stopped dripping, we had reached what soil scientists call the 'full point'. Gravity has pulled as much water as it

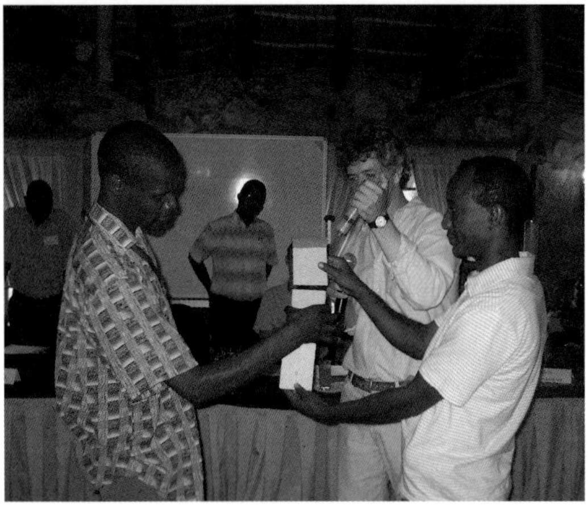

Soil holds water like a sponge. Water is absorbed by a dry sponge, and then moves downwards as each layer of the sponge is filled to its capacity. Soils vary enormously in the amount of water they can absorb. This has obvious implications for how often we should water, and how much water to apply at one time.

can against the attraction forces of the cloth. At the full point, a soil is holding as much water as it can against the pull of gravity.

The first gentle squeeze of the cloth gives us the water easily accessible to plants. For the second tighter squeeze, the plant must work much harder and will start to grow slower. During the white-knuckle squeeze, the plant will use water slowly and start to show signs of wilting.

When we talk about sandy, loamy or clayey soils, we are describing the proportion of different particle sizes. All soils have some sand, silt and clay, but the amount of each varies enormously. Soils containing a lot of sand tend to have large pores which drain under gravity, so sandy soils do not hold much water for plants. Water sticks tightly to clay because of its enormous surface area and small pores. Clay soils hold more water than sandy soils, but once clay soils have dried a bit, plants have difficulty in sucking any more water out because water sticks tightly to the clay surface and in the tiny pores between the clay particles.

We can squeeze a damp cloth as hard as we like, but there comes a time when we cannot wring another drop out of it. In soil science we call this the lower limit, because beyond this point the plant cannot get any more water out of soil. But the cloth still feels damp. Hang it out in the sun to dry and it will lose at least another 20 ml of water.

If we take a bone-dry cloth that has dried stiff in the sun and dip a lower corner into the bath, we see it draws the water upwards. This shows that gravity is not the only force moving water around the soil. In fact water is always moving in the soil, either up or down. Sometimes it's fast and sometimes it's very slow. Gravity pulls water down and the attraction force of drier particles pulls water from wet places to dry places. When water moves from wet to dry, we call this capillary movement, or 'wicking', like a candle-wick that draws melted wax upwards towards the flame.

This explains, for example, the patterns we see around drip emitters, both on the surface and when we dig a cross section. When we drip water on dry soil, the water moves equally in all directions, because the 'wicking' of dry soil is stronger than gravity. A cross section would show the wetting pattern as a semi-circle. After a while the water goes deeper than it goes wide. The reason is that the dry soil is now much further from the drip emitter, and it can't exert as much wicking effect from that distance. Meanwhile the soil just under the dripper is getting even wetter, so gravity takes over and pulls the water downwards. The wetting pattern elongates.

What does all this mean for the irrigator? The soil is considered full of water when the large pores are drained by gravity, i.e. when the dripping from the base of the facecloth has slowed to the occasional drop. The stress point is reached somewhere during the second squeeze, and the wilting point is reached when the plant cannot get any more water out of the soil.

For maximum yields we must replenish the store of water in the soil before we reach the stress point. We must stop irrigating when the root zone soil has reached the full point, because any extra water will drain below roots and be wasted. We need a sensor that measures the full point, the stress point and the wilting point, and all the levels in between, something like the thermostat that runs the central heating of a building. If the temperature falls to a set point – in our case the stress point – the heating is turned on. When the temperature reaches what we desire, or our full point, the heating is turned off.

We can start our search for water sensors by scouring the shelves of the irrigation section of the biggest hardware store in town. We will find all kinds of drippers and sprinklers, electronic solenoid valves, timers and irrigation controllers, but we will be lucky to find an equivalent to the thermometer for soil water measurement. We will see sensors that detect rain and override irrigation controllers. But the closest to a soil water sensor we are likely to find is a little spike with a dial on top that reads from wet to dry, designed primarily for pot plants.

You push the spike into the soil and the arrow on the meter flicks across to record the moisture level. The device operates by having two conducting bits of metal at the tip separated by an insulator. When the probe is pushed into the soil, the soil becomes a pathway or bridge between the two conductors. The wetter the soil is the more current will flow between the conductors, and the further the arrow will swing across the scale.

This instrument does not actually measure the water level in the soil; it measures the ability of current to flow through the soil, which in turn is affected by the amount of water in the soil. But the amount of current that flows is also affected by the amount of sand or clay, the compactness of the soil, the fertiliser level (salt) and the soil temperature.

If you put this instrument into very pure water it will record the water as dry, because pure water is not a good conductor. If you put it into drier soil that contains a lot of nutrients it may read wet, because the nutrient molecules dissolved in the water carry the current. This

The tip of the tensiometer is porous and buried in the root zone. As the soil dries, it sucks a little water out the porous tip causing a partial vacuum inside the gauge. The lettuce roots have to suck water out of the soil, and the reading on the dial tells us how easy or hard this will be.

illustrates the problem we have measuring water in the soil; there are so many other things about soil that interfere with the measurement.

The only way to be sure of how much water is in the soil is to dig up some soil, weigh it, put it in an oven at over 100°C overnight and weigh it again. The change in weight is equal to the water evaporated from the soil. All soil scientists have done this many thousands of times. But this is impossibly tedious for an irrigator.

The most accurate device for measuring the amount of water in the soil is called a neutron probe. You need to drill deep holes into the soil and lower a neutron-emitting source down the hole. Apart from the fact that it is incredibly laborious to get the measurements, you need to do a radiation safety course before you are allowed to start. Many scientists still prefer this method, but it is never going to be easy to use.

My favourite instrument for measuring what's going on in the soil is called a tensiometer. It was invented over 70 years ago, and is still widely used today. The tensiometer does not measure the amount of water in the soil, but the amount of suck a plant needs to exert to get the water out of the soil, which is more useful. Regardless of whether the soil is sandy or clayey, I can use a tensiometer to ensure the plant does not drift into the stress zone where it is getting too hard to suck water out of the soil.

The tensiometer is a water-filled tube, sealed at one end, but with a clay ceramic tip at the other with very tiny pores in it. The ceramic tip is pushed into the soil. As the dry air evaporates water from leaves, the leaves suck water from the roots. The roots suck water from the soil and the soil sucks a little water through the pores in the ceramic tip of the tensiometer. This sets up a partial vacuum inside the water-filled tube, which we measure with a gauge. In other words, the gauge tells us how much suction the roots must generate to get water out of the soil.

A whole lot of new soil water measuring devices have appeared on the market over the last 20 years. Most of these rely on the fact that water is a molecule with a positive and negative side, and so affects the way electrical pulses are propagated through the soil. These devices are usually connected up to a logger which records the amount of water in the soil every hour or so. The farmer can connect up their laptop and download the last week's data. Some systems can dump the data onto the web, so you do not even have to leave the office.

These systems can be a bit pricey, but most work well, and government agencies run training courses and provide financial incentives to get more irrigators to use such equipment. Yet three of every four commercial irrigators in Australia still do not use any of the available devices. There are all kinds of reasons for this. Some irrigators are put off by the expense or time involved in sorting through the data, and others feel their local knowledge and experience is just as good.

You will find a lot of people who talk about saving water, but few who actually measure what is going on in the soil. The expansion and contraction of mercury in the bulb of a thermometer is understood so precisely that no one would quibble with the readings, but we do not have such an instrument for soil water. Those who do measure soil water will tell you that they learned an enormous amount, but it wasn't easy. It remains a great puzzle that there is no straightforward way to get a better grip on water in the soil. These days there seems to be an inexpensive gizmo to measure everything, so why not the most important thing of all?

Chapter 10

The Machingalana is talking to me

My first job growing lettuce on the university research farm required me to use as little water as possible. Despite completing a four-year degree in agriculture, I felt ill-equipped for such a practical assignment. I asked the local horticultural advisor to put me in touch with the best lettuce grower in the district so I could learn the ropes.

Sitting around his kitchen table, Sam the lettuce grower explained how he irrigated his crop from sowing to harvest during cool, warm and hot weather. His unit of measurement was hours his sprinklers ran, which was not very useful for my drip-irrigated crop. The description of big and small lettuce together with hot and cold days was far too vague for me to derive any benefit from the conversation. Sam was a good lettuce grower, but for him it was an art, not a science.

I turned to the soil water monitoring tools, and they became a lifeline to me. I found the soil was often far drier or wetter than I thought, and I wondered how I ever irrigated properly without these aids. But there were problems.

The neutron probe is a device where a radioactive source is lowered down a tube, which penetrates over a metre deep into the soil. The technique gives accurate measurements of the proportion of the pore space occupied with water, but lugging the heavy instrument around the field and manually recording many depths in many holes took up half the day. Should I spend several hours taking measurements, or just irrigate anyway and get on with some other pressing task?

I installed tensiometers which show how hard the plant must suck to get water out of the soil, and the first job each day was to walk up and down the rows reading the values on 30 or so gauges. Tensiometers do not show the water status of the soil dropping uniformly like, for example, the petrol gauge in a car on a long journey. The tensiometer may take three days to go from the full point to three-quarters full, and then read 'too dry' the next day. This is how soil works. At first it's easy for the plant to suck water out, but it quickly gets harder and harder, and you can easily be taken by surprise. No matter how hard I tried, the tensiometers recorded some parts of the field too wet and others too dry.

My first project in rural Africa was at an agricultural college, training students to be extension workers for the small-scale farming sector. The college farm was fed by a dam in the nearby hills and the water flowed past in a small concrete-lined canal. The farm used what it needed and then the canal carried the remainder to the next village or two before the supply petered out. We at the college got first use out of the canal, and what happened downstream was someone else's concern.

The canal filled a holding dam on the farm, from which water was pumped to lines of sprinklers irrigating maize. Some time back, no one could quite remember when, instructions were laid down as to how long each line should be left to irrigate. At any one time the sprinklers covered about 20% of the area to be watered. Someone had to turn the pumps on and off and had the laborious job of moving the pipes every so often, so the whole area could be irrigated in rotation.

Once again I was doing research about giving exactly the right amount of water to the crops. The less water we used, the more would flow down the canal to the next village, and that was only fair. We installed a pan of water, just over a metre in diameter, which was refilled to exactly the same depth each morning, so we could see how much water evaporated the previous day. We used this evaporation number, combined with a correction factor which was increased as the plants grew, to estimate how much water the crop would require each day. We knew the rate at which our sprinkler applied water, so we could work out the irrigation run time for each plot.

This is called the evaporation pan method, and is taught to every student of irrigation. It involves one simple measurement and a couple of simple calculations. It is not dead accurate by any means, but it keeps the water applied within reasonable limits.

From our research plots, I watched the sprinklers on the maize pumping out water all day long. The more water the manager put on each place, the longer it would last until the next irrigation, the less he would have to worry about plants running out of water and the fewer times he would have to move the pipes. He may be wasting water, but the strategy made some sense, except for the villages further down the canal.

At field days in Australia I demonstrated the evaporation pan method, neutron probes and tensiometers to farmers. The vegetable farmers nodded politely, but there were few, if any takers. I did the same in Africa, but this time my audience were small-scale farmers and their advisors. One hot day under the African sun, while addressing a group of farmers, the thought suddenly struck. Why would small-scale farmers adopt equipment that I couldn't even get the big operators in Australia to use? It slowly dawned on me that we had little to offer these people.

My next project in Australia involved working with some of the new fancy water monitoring gear that had just become available. These were soil probes, connected to a computer or logger, and they measured soil water content every 15 minutes so we could see what the plant was experiencing all the time. It was a revolution. No more time-consuming manual readings, and wondering what was happening between them. We could see what the plants experienced; the soil wetting up during irrigation and drying down in between, the difference between warm days and cool days and even day and night.

I needed to install the equipment at an experimental site a few hundred kilometres away. To make sure that the gear was working correctly, I installed these new soil probes in my vegetable garden at home. I had 16 probes, which were placed at two depths in eight different garden beds, and then wired up to a computer. Just when I had established that all the equipment was working well, I got news that the experiment I was preparing for had to be abandoned. So I just left the probes measuring away in the garden. Now I would have the most efficiently irrigated vegetables in town.

Each week I downloaded the data from the computer, but to my surprise even all this information did not completely solve my irrigation dilemma. The so-called full point and stress point seemed to be different for each bed. When rain fell, the instrument showed the soil water level increased, but by a different amount in each bed. Measuring water was one thing, but what to do with the measurements was another.

When I watered for half an hour, the instrument recorded the soil at the full point. But if I kept watering, the soil water content did not go up further because the excess water drained through. It is a bit like dripping water onto a fully wet sponge. Water drips in the top and then drips out the bottom at the same rate, so we are wasting water but the weight of the sponge (its water content) stays the same.

At first I thought it was marvellous to have water measured in 16 places at 15 minute intervals night and day, but soon it became information overload. All I really wanted to know was if I had applied enough water. If I had young lettuce plants, I wanted to know that the water did not go past 15 cm, because their roots were shallow. If I had fully-grown tomatoes I wanted the water to go much deeper.

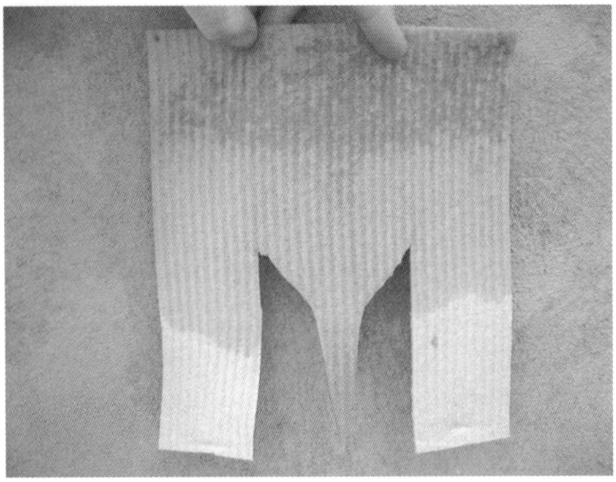

It is hard for irrigators to visualise what is happening down in the root zone below the soil surface. This sponge shows a wetting front moving downwards. The funnel shape cut into the sponge converges the moving water towards the tip. Drops of water drip from the tip of the sponge. This simple idea started me on a long journey of invention.

If just knowing whether water did or did not reach a certain depth could help me to be a better irrigator, there could be a simple way to get this information. I grabbed a funnel from my young daughter's sandpit, stuffed toilet paper in the neck to make a crude filter, and filled the funnel with soil. I dug a hole in the garden bed and placed the soil-filled funnel in it with the narrow end jammed in a glass jar sitting at the bottom of the hole. I filled the hole in and watered the garden bed. Next day I dug it up, and the little jar at the bottom of the hole was filled with water.

This was the beginning of the wetting front detector project, which was to dominate my thinking on irrigation for quite some time. A work colleague and I built a number of prototype versions, all based on the idea of a buried funnel. We couldn't dig it up to see if had collected water, so we used an electronic float switch in a small reservoir at the base of the funnel behind a ceramic filter. When the water reached the reservoir, the switch activated a light or a buzzer.

The wetting front is the sharp change in soil water content that occurs during irrigation. The front is essentially the divide between the wet soil above and the drier soil below. Wet soil is a darker colour than dry soil, so the wetting front is usually quite easy to see.

If we apply small amounts of water frequently with a hose or sprinkler, then the surface of the soil will be kept moist. Much of the water will simply evaporate from the soil surface and not be of any use to the plant. If we apply large amounts of water once in a while, then the water may infiltrate below the roots of the plants and be wasted. Most irrigators have some idea how deep they want the water to go into the soil, and this is what makes tracking wetting fronts useful.

During irrigation, the soil inside the funnel became super-wet, a bit like a magnifying glass focusing a beam of light. As the water moved down towards the narrow end of the funnel, the water was channelled into less and less soil. At the base of the funnel the soil got so wet that water seeped out of it, passed through a filter, and dropped into a small reservoir.

In the early days we connected a solenoid valve to the electronic switch at the base of the funnel. A controller turned the water on as per usual. But instead of the controller turning the water off after half an hour or so, we let the float switch do it. In this way we could be sure that the water had infiltrated past the depth our new instrument was buried.

We had to find a name for the new device and settled on 'FullStop Wetting Front Detector'. By 'FullStop' we meant 'when the soil is full – stop irrigating'. This name had its roots in the very first experiments we carried out to test the electronic version of the device. We buried the detector under turfgrass and had the sprinklers come on automatically every fourth day. When the infiltrating water reached the detectors they automatically switched the water off.

If the weather over the past four days had been hot and dry, then the soil would be quite dry. When the sprinklers were turned on it took some time to fill up all the empty pore spaces. The wetting front moved slowly and a long irrigation was allowed by the detector.

If the weather had been cool, or there had been some rain, the soil would still be quite wet on the irrigation day. The wetting front now moved quickly through the soil because most of the soil pores already contained water. In this case the wetting front reached the detectors after a short time and a corresponding small amount of water was applied. In each case irrigation was stopped when the top part of the soil was full of water – FullStop.

I now had a device that we could test back among the small-scale farmers in Africa. We replaced the electronic float switch with thin rods of styrofoam. The styrofoam floated on the water in the reservoir and in turn made an indicator pop up at the ground surface.

My students wanted to find a suitable African name for the device and came up with 'Machingalana' (ma-ching-ga-laana). A Machingalana is roughly translated as a watchman or security guard. It is common in South Africa when entering a property to pass a security guard and sign your name in a book. The guard or Machingalana might also operate a boom gate at the entrance with a STOP sign on it. My student visualised the FullStop Wetting Front Detector as a Machingalana. If the water wanted to go past it had to sign in on the way, just like everybody else.

One of the first places we took the Machingalana was to the group of small-scale irrigators nearby the agricultural college which had started this journey several years before. Most of the farmers there had small food plots growing vegetables. We explained the principle of a wetting front detector and found a farmer who was willing to work with us. We installed two detectors in his plot, one at a shallow depth and the second a bit deeper. We left the farmer with a recording sheet, and asked him to mark the days he irrigated, and whether the shallow and deep detectors responded to the irrigation.

It turned out that our chosen farmer could not read so he enlisted an old man, who had been a school teacher, to help him fill in his form. The old man was also a farmer, and he was

The wetting front detector catches some of the infiltrating water in the funnel. The water seeps through a filter at the base of the funnel and is collected in a small reservoir. This causes styrofoam rods to float, thus activating a magnetically latched indicator above the soil surface.

intrigued by the Machingalana. The old man was growing wheat, and he asked our extension officer if he too could get a Machingalana or two for his wheat field.

When our extension officer visited the old man, he saw that the young wheat crop was already deficient in nitrogen. The old man had used all his money to plant the crop and had therefore cut back on fertiliser. Our extension worker estimated the potential yield of the young crop, and it seemed as if the farmer would be lucky to recoup his costs. He would have to apply some fertiliser, and this is what he decided to do.

Shortly after applying nitrogen fertiliser, our extension officer received the message that rain following irrigation had set off the deep Machingalana. The old man was worried about leaching all his fertiliser out of these sandy soils. He was allocated two irrigation shifts a week, but decided to skip some, to ensure his shallow Machingalana responded, but not the deeper one. The old man takes up the story:

During 2001 some of our farmers were fortunate to receive five hectares (ha) of land. I have decided to plant wheat, but because I am afraid of debt, I used my saving money for this exercise. I have prepared 2.5 ha before the time and have planted during June 2001 and harvested during November 2001. I am aware of the fact that I have planted at a seed rate below the recommendation and that the top-dressing was about half that is generally recommended.

The Machingalana helped me to make optimum use of the nitrogen applied on the wheat crop as a top-dressing. Without managing the irrigation application according to the principles of the Machingalana, I would have wasted most of the nitrogen application into the Elands River and would have battled to average a yield of 5.4 tonnes per hectare (t/ha). This became even more significant when I have compared my results with some of my fellow farmers who were not participating in this project, and who have harvested 2.4 t/ha on average. I have experienced that this Machingalana was working – it was talking to me!

I doubt we can ascribe this doubling of yield to the Machingalana. The irrigation system was full of leaks and no two nozzles looked the same, so it would be difficult to make definitive conclusions from the two instruments buried in his field. No doubt it was the extra fertiliser that made most of the difference. Yet the story shows that simple tools can give farmers the confidence to change their practices and improve their yields.

Over the next three years we worked to get our home-made prototype detector commercialised. This meant finding a partner who could fabricate the moulds to make the plastic parts needed to build a modular detector that could be packed in a box for sale and then assembled in the field. Although we aimed to reach the small-scale farmers, it was the large-scale end of the market that picked up the new device. At last count over 13 000 of these detectors had been sold around the world.

A wetting front detector has its limitations, but it did achieve two important aims. First, how deep the water goes is the way many irrigators think about what they are doing. Most do not think in terms of the suction a plant must exert to pull water from the soil or the proportion of the pore space occupied by water. But they do understand that shallow irrigation means that most of the water is lost by evaporation and very deep irrigation may be wasting water below the deepest roots.

Second, the wetting front detector collected a sample of water, and some farmers started measuring the nutrient and salt contents of the water captured by the detector. As we will see in later chapters, it is impossible to be a good irrigator without measuring the nutrients and salts carried in the soil water.

Sometimes I think how far I have fallen, from the fancy equipment I worked with that measured the precise water content every 15 minutes, to a plastic funnel with a float on top. But then I have a vision of the old man resting from the hot sun under an acacia tree and discussing the finer points of irrigation and nitrate leaching with his illiterate friend.

Chapter 11

Watering grass

A once common sound is disappearing from our neighbourhood. On any Saturday morning you could hear several lawn mowers, some roaring nearby and the distant drone of those further away. These days the lawn is the bad guy of the urban water crisis. We are encouraged to save water inside the house, there are rules about when we can water outside, but current water restrictions mean it is forbidden to water the lawn. The lawn is seen as profligate and superfluous.

It is true that around 40% of domestic water is used outside the house and that much of the average garden is taken up by lawn. Some lawns are left to fend for themselves – they go brown in the dry times and revive after rain – but many get the haphazard sprinkler treatment. The sprinkler is left to sit in one spot until someone remembers to drag it to the next.

Pop-up sprinklers connected to an irrigation timer make the job much easier because they are designed to apply water evenly over the whole lawn at regular intervals. Pop-ups were becoming very popular, and produced the lushest lawns of all, but have now fallen on hard times.

Several years ago, our garden was opened to the public as part of the Australian Open Garden Scheme. It is at events like these you realise how many avid gardeners there are out there, and how keen they are to learn about water. I had just given a formal seminar in my place of work on irrigation to a rather disappointing turn-up. Now the paying public were rolling up to our garden by the hundreds. I gave talks on water and the audience were crowded on the veranda and along the garden paths. Some even perched in the trees.

The next Monday morning I received an email from an Open Garden visitor asking me whether I could help him to trim his water bill, which had topped a million dollars. He worked for the part of local government responsible for the irrigation of sports ovals across the city.

In the agricultural sector, Australia pours more water onto grass than onto rice, cotton, fruit and vegetables combined, largely to grow pasture for dairy cows. Grass in the city, be it the domestic lawn, the sports oval, parks and gardens and most of all golf courses, demands a significant slice of the urban water supply. From both agricultural and urban perspectives, it is worth putting in the effort to irrigate grass as well as we possibly can.

I am used to travelling to field sites in far away places, but my new sports oval field site was at the end of the street, where the kids played their weekend sport. My brief was to work out how much water could be saved on sports ovals, and it looked to be a straightforward assignment. Frequently mowed turf has the same amount of leaf exposed to the sun all year, so all we need to do is adjust the water for the prevailing weather.

I was pleasantly surprised by the sophistication and efficiency of sports field irrigation across the city; only two people were responsible for irrigating over 100 ovals. Some time back, an in-house research team had carried out a series of experiments to work out the exact water requirements for turf in the city. Then an automatic control system had been set up that could open and close the hundreds of water valves that fed the underground pipes to each oval, all by radio signal.

The system was also a fault reporter. Each valve had to operate within certain flow limits. If the flow rate was too high on a particular line, the valve was automatically shut down on the assumption there was a broken sprinkler. If the flow rate was too low, then a blocked nozzle was flagged. Each morning the system generated a giant print-out of what it had done the previous day, the amount of water applied to each oval and the list of faults it had detected.

The irrigation system also calculated the amount of water the grass could transpire each day. The method for calculating grass transpiration is based on evaporation from an open pan. This is a simple and robust method that has been used for many decades around the world. It is not usually automated to such a degree, but the basic principles are well accepted.

The typical 'pan' is 1.2 m across and 25 cm high and filled to within 5 cm of the rim. Each day the pan is refilled to exactly the same point, so we can measure how much water evaporated the day before. Pans are typically measured at 9 in the morning, and yesterday's evaporation figures are published in the daily newspaper and posted on the web.

We normally measure the amount of liquid by volume, but it is far easier to measure rain and evaporation by depth, usually in millimetres. If 10 mm of rain fell overnight on a flat car park with curbs but no drains, then the impervious surface would be covered to a depth of 10 mm in the morning. When we measure 10 mm of rain in the rain gauge, the depth of water looks much greater, because the rain gauges do not have straight sides. The catching end has a greater cross sectional area than the base, which makes it more accurate to read small amounts of rain.

The evaporation pan is a way of estimating how much water the grass could use. The depth of water evaporated is measured each day. Canberra ovals get 65% of what evaporated from the pan, dished out two or three times a week.

Evaporation is much greater than rainfall across most of the country, which is why we need to water the sports ovals. Canberra gets about 600 mm of rain in a year and the pan loses about 1800 mm through evaporation. The water level in a swimming pool would fall by about 1200 mm or 1.2 m in an average year if it was never topped up.

The amount of water used by the grass and the amount that evaporates from the pan are related – both the pan and the grass lose more water on a hot day than a cool day. It turns out that a ratio of around 0.65 keeps the grass looking pretty lush. So, if 10 mm evaporates from the pan, then we need to apply 6.5 mm to the grass.

We could irrigate the grass every day by applying 65% of the water lost from the pan the day before. Watering every day is not an efficient way to do things because each time we turn the sprinklers on some of the water evaporates before it hits the ground, or evaporates from the surface of the leaves. It is best to add up several days of water use, and then apply it at one time.

But how many days can we add up? The grass might not last a week without water. Or, if we added a lot of water occasionally, then the water may seep down below the deepest roots. To get the interval right we need to know something about how much water the soil can store and how deep the roots go. So the automatic control system is told what type of soil is on every oval. It is also told that the grass roots are all growing within the top 100 mm of soil. If water infiltrates deeper than 100 mm, the system believes that the grass can't use it, because there are no roots down there.

Now we are ready to run the system. Let's say the evaporation from the pan over the last three days was 6 mm, 4 mm and 5 mm giving a total of 15 mm. The grass water use is 65% of that, or about 10 mm. It is time to apply 10 mm of irrigation to fill up the soil store again. This simple calculation goes on day after day on 100 ovals, with valves opening all over the city when the grass has emptied about half of its store, and closing when the soil store has been refilled.

It is a very ingenious system until the rules of the game change, and this happened when water restrictions came in. At first it was mandatory to reduce water use by 15%. But how do you save 15% if you are already operating at peak efficiency? Local government made the obvious decision; they stopped watering 15% of their ovals. But they also went a step further. It was time to check whether their automatic system was as efficient as everybody had come to believe.

The first step of the investigation was to measure how much water was in the soil each day. I installed five soil moisture sensors at four metre intervals between two adjacent sprinklers. The sensors were connected up to a logger which recorded how much water was stored in the soil every 15 minutes. A month later I downloaded the logger, to find that the water content at each location between the two sprinklers was very different. Two frequently showed the soil as very dry, two as wet, with one in between.

This is the first practical problem we have with trying to measure how well we are irrigating. If, for example, we needed to know the temperature, we would only need one or two thermometers to cover the city. There is never that much disagreement about how warm or cold the weather was today. But five soil moisture sensors covering 2% of the oval gave three different opinions. The differences in soil moisture can be traced to four sources of variability: variability in the soil, in the grass, in the uniformity of irrigation and in the way the sensors were installed.

The sensors were all within 20 m of each other, and the grass was mown twice a week, so variability in soil type and grass leaf area would be small. Installation is always a problem. The very act of getting a sensor into the ground changes the soil around it. In this case I suspected that the uniformity of sprinklers was the main culprit.

The uniformity of sprinklers is easy to check. You lay out 20 containers in a grid pattern between three adjacent sprinklers. After irrigation, the amount of water in each container is measured, and the average calculated. Then you take the five containers with the lowest amount of water in them and compute their average. Say the average in all containers was 10 mm, but the average in the lowest five containers was 6 mm. This gives us a Distribution Uniformity (DU) of 60% – 6 mm divided by 10 mm.

Now if I aim to apply 10 mm of water to the entire oval, 25% of the area will get 6 mm (five containers out of 20). Some areas will get more than 10 mm. If every part of the oval needs at least 10 mm to stay lush, then I have to over-water most of the oval so the driest quarter gets its 10 mm. A properly designed sprinkler system should give a uniformity of at least 75%. Most don't; my oval was in the mid-60s.

I couldn't solve the uniformity problem. You can change the type of nozzle on the sprinkler, which had been done, but you can't make the underground pipes bigger, or get them closer together.

We buried wetting front detectors throughout the oval to record when the water went past the shallow roots of the grass. Irrigation was never supposed go below 150 mm – if it did too much water had been applied. After the change to recycled water, we measured the build up of salt from the water sample captured by the wetting front detectors.

Despite the sophistication of the irrigation control system, I knew that there was plenty of water to be saved. I buried five wetting front detectors in a line between two sprinklers, to catch any water moving past 150 mm depth – a bit deeper than the 100 mm rooting depth limit programmed into the system. If the irrigation system was accurate, then irrigation water should not infiltrate deeper than 100 mm and the detectors should never collect any water. They might respond to rain, because we can't turn the rain off. But if they responded to irrigation, which they frequently did, then one or more of the settings in the control system was incorrect.

The first input to check was the evaporation pan data, because this was the number driving all subsequent water use calculations. The pan measuring evaporation is situated at the airport, in an open and fairly windy place. Most ovals are in much more protected places and therefore in a moister microclimate. I compared the evaporation pan at the airport to another one elsewhere in the city and the airport pan lost 17% more water over the summer.

The next check was to see if roots did really stop at 100 mm depth. I installed more moisture sensors to measure the soil water content between 100 and 300 mm depth. Sure enough, there were roots taking up water from this depth zone.

The problem of getting the rooting depth wrong is that we do not make good use of the rain. If we tell the computer that the soil can only store 15 mm of rain over the 100 mm depth of soil and we get 30 mm of rain, then the computer assumes the other 15 mm will drain below the root zone and be wasted. The reason the detector collected water was that the computer ordered an irrigation when the grass, with access to deeper water, did not need it. Assuming a very shallow rooting depth also meant that the oval was irrigated too little too often, which increased the proportion of water that evaporated from the soil surface.

Heavy users of domestic water in Canberra are charged over $2.50 for every one thousand litres. Assuming this applies to the government as well, then 1 mm of rain over one hectare of grass, the area taken up by an oval, is worth over $25. Our annual rainfall of 600 mm over 100 ovals is worth $1.5 million, so it's worth using it as well as possible.

As the drought carried on, the oval was converted to treated recycled water. The recycled water contained a little salt, and this provided us with another view on how well the turf was irrigated. Plants use only the water, leaving the salt behind in the soil. If the oval was given too much water then this salt would be washed through the soil. If too little, the salt would accumulate to levels that would harm the grass. Good irrigation should allow the salt to build up because this means that we are not needlessly flushing water through the soil. Then, when the salt has built up to a threshold, we can apply extra water to move it deeper into the soil.

I reported all this back to the oval managers. I thought the soil water content measurements taken every 15 minutes in five places over a whole season would impress them. But there was a problem. I was measuring one small part of one oval and they had to manage 100 ovals. They simply did not have the time to study reams of data and then interrupt their automatic system to top up one oval today and cut back on another tomorrow.

The experience taught me the difference between giving tactical and strategic advice. The coach of the rugby team sends the team into the game with a strategy, based on their strengths and the opponent's weakness. At half time, in response to the way the game is unfolding, they talk tactics. Tactics are what you have to do next in response to the immediate situation, when the strategy isn't going as well as you thought.

I had wanted to give the irrigators tactical information of soil water content – 10% more water on this oval today, 20% less on that one tomorrow based on my measurements. But what they needed most was to improve their strategy. Strategy is about getting a better understanding of the big picture so you make more robust plans.

The irrigation strategy was based around evaporation from a pan at the airport and the amount of water the soil could hold, based on soil type and rooting depth. It was clear to all that the strategy could be improved, because there were certain times of year when the detectors collected water regularly after irrigation. Nobody could argue with that. The new irrigation strategy needed to increase the rooting depth of the grass in the code of computer program. Measuring evaporation from several different places around the city, or better still automatic weather stations, would improve things still further. This would improve the estimate of how much water should go onto every oval every day, and reduce the need for tactical adjustments.

The interplay between strategy and tactics is important. If we think our strategy is so good, we do not take the measurements that can show us it could be improved. If we just measure, we find we are responding to the recent past all the time, never putting in place the structures to manage the system better. In the case of the oval, we had to start measuring to evaluate and improve the strategy. From time to time we will also have to make tactical adjustments for new and unforeseen events.

As I watch the under 13s rugby on the weekend, it is nice to know how much effort has gone into this surface they are playing on. They run unawares over all my gear buried just a few centimetres below their feet. There was a while when the dam levels got so low that it looked like many ovals would have to be closed. But with childhood obesity now recognised as a major problem, it will be cheaper to find new sources of water and do some monitoring to check things are on track, than to pay the future health costs of kids living on junkfood in front of little square screens.

This summer the detectors did not see much water go past. The water they did collect had a salt level 10 times higher than the salt in the irrigation water. This tells me that very little water has been going past the roots. If it had been, it would have carried the salt deep into the soil. The kids still throw themselves around on the sports field, and so far they have not noticed the oval is getting by on the least possible water.

Chapter 12

Grey water and the washing machine

I hooked up the bath and shower outlets to the garden long before the 2000 drought. There is something deeply satisfying about using the bath water on the garden. No longer do you have to get the kids to wash in 10 cm of water. There are no water restrictions inside, so I am secretly pleased to see a nice full bath after a long hot day, and to then run it out onto the garden.

We have the kind of house where I can crawl under and meddle with the pipes. Most people can't do this. Their easiest source of waste water is from the washing machine, but the water coming from the washing machine is different from the water coming out of the bath. Bath water contains a bit of soap and shampoo. Each load of the washing goes into the machine with a big scoop of powder laden with salt, and all this salt will end up in the root zone of my plants.

Salt is not good for plants, so there was no way I would add it unnecessarily to my soil. Many people asked me during the drought whether they should use washing machine water on the garden and my advice was: 'bath and shower, yes – washing machine, no'.

Once, on a long-haul flight, I sat next to a plastics engineer. It turned out that my fellow traveller was making the extension hoses for washing machines for people to irrigate their gardens. Since we had 13 hours on the plane ahead of us, I told him about my work monitoring salt around the country, and the perils of using salty water for irrigation. I told him about the wetting front detector, and how we had developed an instrument for easy monitoring of salt build-up in the soil. I was likely to be the most ardent critic of the washing machine extension hose on the whole plane, probably in the whole country.

My travelling companion was certainly aware of the potential salt problem but he had a more pressing problem right then. The demand for washing machine extension hoses in Australia was outstripping supply. His factory was working day and night. He may even have to get his factories abroad currently making hoses for swimming pools to switch over to making washing machine extension hoses, just to satisfy the Australian demand.

A few weeks later, in early spring, I walked into our local hardware store, a giant warehouse with 50 aisles. In the foyer were boxes of washing machine extension hoses on display and round the corner there were more. In fact one whole wall contained boxes of hoses stacked from floor to ceiling. I would not have believed there were that many washing machines in the entire city. Evidently, watering the lawn from the washing machine, unheard of a year or two back, was becoming very popular indeed.

I was recounting this story at a formal dinner to a table of distinguished people in the water industry, and again voiced my opposition to using washing machine water on the lawn. 'Well I do it,' answered one of the dinner guests, 'and the lawn looks fantastic'. I gave the usual environmental scientist retort, 'things always look OK in the short term but it's the longer term changes we have to worry about'. I went on to explain how sodium in washing powder makes clay soils less permeable, leading eventually to waterlogging and death of the lawn. 'Funny you should say that,' answered my fellow dinner guest. 'The washing machine water infiltrates into the lawn faster than the town water'.

I was beginning to feel on less secure ground. Maybe the detergent in the washing powder was acting as a kind of wetting agent aiding infiltration. Then the next eminent dinner guest weighed in. 'On our farm we've used all the waste water from the house on the garden for decades with no ill effect'. Clearly, I was going to have to look much deeper into the issue of washing machine water before I told this story again. So I retired to the laundry.

The box of laundry powder advertised 'just one scoop for outstanding brilliance', with the rider, 'measure the correct amount … using more than the recommended amount will not improve results'. Just one small scoop turned out to be 45 grams. I measured the salt content of the wash water and the rinse water with my pocket salinity meter.

By the time the whole cycle was finished I had a barrel containing 150 litres of water, with a salinity level far lower than I expected. I sent the samples down to our labs in Adelaide 600 km away at the other end of the river catchment to check. Sure enough, the water that came out of my washing machine had less salt than their drinking water. It was a hot and windy day outside and a day I was not allowed to use mains supply. So I rounded up all the washing I could find, did three more loads and ran the water out onto the garden. Plenty of irrigators I knew along the River Murray used this quality water for their crops. I was converted.

Several months later, and I was not so sure. The soil where I had used the most laundry water seemed to have developed a hard crust and some of the crops looked to be a little below par. There were suspicious yellow mottles on the apple tree leaves. My neighbour, following my example, had turned his blue hydrangea flowers to pink. I measured the pH of the washing machine water and it was very alkaline. Maybe the salt was affecting the soil structure and maybe the pH of the soil was rising and affecting the availability of some nutrients.

So what's in laundry detergent, and what might be the shorter and longer term impacts of using it on the garden? What are the essential constituents of a good laundry detergent and which of these are compatible with the garden?

The starting point is to understand soap. Soap is a long chain molecule with one end that is attracted to water and the other end attracted to fat and oil. The soap molecule essentially sticks water to fat. In other words soap allows two liquids that normally repel each other – oil and water – to mix, and the oil becomes suspended as tiny globules in the water. Soap also reduces the surface tension of water – the tendency of water to bead when placed on a hard surface. Lowering of the surface tension gives soapy water the ability to penetrate fabrics and get between the fabric and the dirt. These days most washing products have replaced soap with detergent, which is basically a synthetic soap.

I have heard of people who grate up a bit of soap and use it to wash clothes with pretty good results. It all sounds so easy, so why does the box of powder in our laundry says it contains enzymes to break down stains, agents for water softening and suspending dirt, chemicals for bleaching and optical brightening, fabric softeners, perfumes and more. I know we need the detergent, but do we really need all this other stuff, and what is it doing to the garden?

You can wash clothes in just soap or detergent if the water is fairly pure. Many town water supplies contain calcium and magnesium, dissolved from rocks as water percolates through the soil. There is no problem drinking this dissolved calcium and magnesium and it does not affect taste, but in high enough concentration it makes what we called 'hard' water. Soap does not lather up in hard water. The reason is that the calcium in hard water reacts with detergent to make a precipitate we know as scum.

Scum is that brown ring around the bath after the water has been let out and it's a big problem for washing clothes. When calcium in the water reacts with detergent in the washing machine, the detergent becomes ineffective and we get a scum that sticks to the clothes and makes whites look yellowy. So the second ingredient we must add to the box of laundry powder after detergent is a water softener. The water softener binds up the calcium and magnesium, thus preventing their reaction with the detergent.

The third ingredient is the dirt suspender. This helps keep the dirt dissolved in the water so it's easy to wash out in the rinse cycle. We can class the enzymes, bleaches, optical brighteners, fabric softeners and perfumes as added extras. They may be useful in some circumstances, but probably not essential to getting the clothes clean.

So which of these components is a problem in the garden? Essentially there are four basic constituents in laundry powder that we have to think about. First there is sodium, which plants have no use for at all. Sodium gets between clay particles and breaks them up so soil pores become blocked and the soil becomes less permeable. Phosphorus and boron are plant nutrients, but can be toxic to plants at high levels, particularly boron. Finally there is the level of acidity or alkalinity; the pH.

Sodium is part of the detergent molecule itself. Sodium compounds are also used as water softening agents to stop scum formation and to make the water more alkaline because most detergents work better in an alkaline solution. Sodium is part of some anti-microbial agents, bleaches, corrosion inhibitors and agents that stop dirt redepositing on clothes. When it comes to sodium, we have a head-on collision with the needs of the garden. Sodium is fundamental to laundry detergent but not wanted in the garden.

Phosphorus in detergent is used as a water softener and as an agent to suspend dirt for easy rinsing. Unlike sodium, phosphorus is an essential plant nutrient and often deficient in soils. Australian native plants are adapted to low phosphorus soils and can be damaged by high levels of phosphorus, but a fruit and vegetable garden does need a steady supply. These days most washing powders claim 'no phosphorus' among their green credentials. This is because phosphorus contributes to algal blooms in rivers, and phosphorus in the sewers of inland towns is en route to a river.

Boron is a component of some bleaches and fabric softeners. Boron is needed by plants in trace amounts and is sometimes added to soil but can easily reach toxic levels where it will harm plant growth. These days few detergents contain any boron. If boron is really needed for plants it should be added in precise quantities, and this won't happen through the application of waste water.

Lastly we come to the pH of the detergent. All laundry detergents have high pH. Again there are differences in what we call the alkalinity of the different brands, which is the extent to which the detergent increases the pH of the soil.

The labelling of detergents as concentrated, biodegradable and environmental friendly or garden friendly is helpful but not foolproof. A good number of the brands with the 'eco' or 'green' word in the title are low salt, low phosphorus, but this is not universal. Some of the 'no-name' brands are as good as those with fancy names, and some of the best-known names in the laundry business are the saltiest of all. Fortunately, a knowledgeable soil scientist has done all the hard work and put it up on a website for us all (see chapter notes at end). It is worth checking the credentials of your favourite detergent.

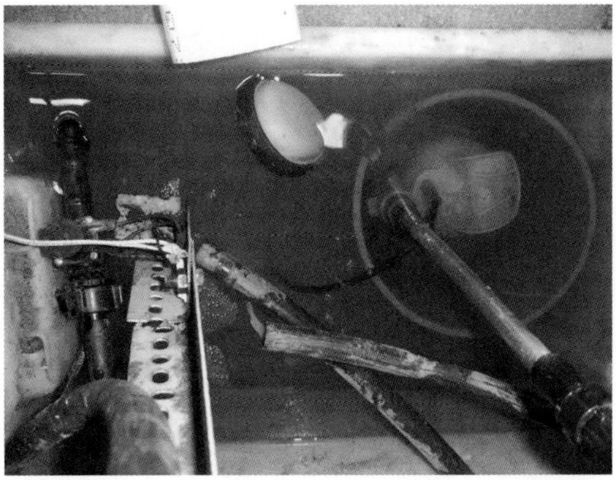

Commercial home grey water systems can cost far more than the value of water they save. We built a simple system using a 400 Watt pump with a float switch. As soon as the bath emptied into the container the grey water was pumped out into the garden.

The biodegradable tag makes less sense for the gardener. A biodegradable poison makes sense because the large 'active' molecule is broken down into its safe constituent parts. Sodium, phosphorus and boron cannot be broken down any further than they already are. No matter how biodegradable the product is, the concentration of these elements does not change. Biodegradable refers mainly to the detergent, which does break down rapidly. Curiously, garden stores are pushing wetting agents to improve watering, which are simply detergents that are not biodegradable.

Plenty of laundry detergents now contain no phosphorus or boron, but they all contain sodium, so we need to find a low sodium brand. Amazingly, some brands have just one-tenth the sodium of others, so there is a lot of room to move. The liquid detergents tend to have much lower sodium than the powders, and many have a neutral pH as well. So if we can purchase low sodium, neutral pH, no phosphorus and boron liquid detergent, will they still get the clothes clean?

Our top loading washing machine died during the preparation of this chapter after 20 years of hard labour. We changed from top loader to front loader (because it uses less water) and from powder to liquid at the same time. The general consensus around here is that the washing is not quite as clean and the clothes a little stiffer. It is bad science, of course, to change two factors at the same time.

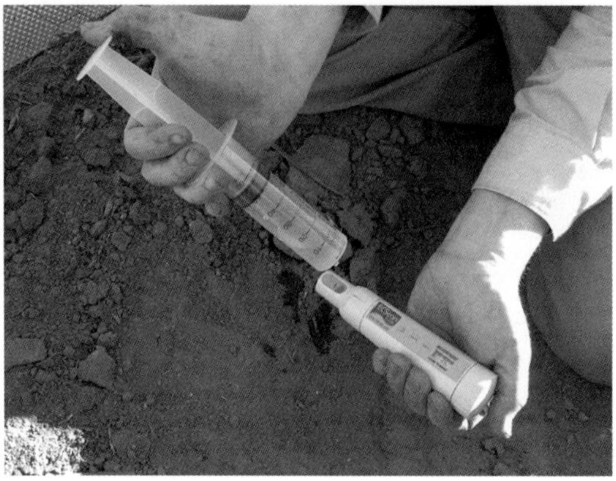
Using grey water means that all the bathroom and laundry products end up in the soil rather than in the sewer. I use wetting front detectors to collect water samples from the soil. The amount of salt in the water is easily measured with this small meter.

The Australian Consumer Association has rated laundry powders for washing effectiveness and suitability for the garden. Unfortunately the general trend shows that powders that gave the best results contained the most salt. There were exceptions, however, and there are brands that are slightly less effective cleaners, but produce waste water 10 times better for the garden. And all we need is one exception to prove the rule that we do not have to have salty, alkaline laundry detergents.

It is one thing to strike a compromise between cleanliness and looking after the soil, but another thing to use laundry water efficiently in the garden. The legal requirements vary around the country, but in general you are supposed to use grey water in a drip system under a considerable depth of mulch, and not on food crops. The aim is to stop people coming into contact with the grey water, and detergent on leaves cannot be good for them either.

Moreover it's not legal to store grey water on site and you would not want to either – it stinks after a remarkably short time. In our simple system, the bath, shower and laundry water gravity feed into a large plastic drum under the deck, containing a 400 Watt pump with float switch. When the shower is turned on the water runs out of the house into the drum. The float switch is activated, pumping the water out onto the garden until the drum is near empty.

It won't be long before there are strict guidelines on the use of grey water. In the meantime we use it on the two very dry back corners of the garden where there are fruit trees, and big thirsty trees on the other side of the fence. On the one side we just let the water run from a hose into a basin around the trees. On the other we have a plastic-lined walled bed about 60 cm high and dug 30 cm into the soil. A perforated pipe is coiled at the base, covered with bark chips, then 30 cm of sand, then soil. The grey water is fed into the perforated pipe and slowly percolates upwards and then can overflow about 30 cm from the surface. Pumpkins or some such rampant crop are grown above, and the overflow water trickles past the fig and avocado trees and down past the raspberries and, if it makes it, joins up with the soakaway trench.

Having spent some time researching laundry detergents, I have currently disconnected the washing machine from my waste water system. My problem is the wads of fluff that block the filter into my pump. Evidently our clothes are slowly disintegrating in the washing machines. I still delight in using the bath and shower water, which play a major role in keeping the garden productive in summer. In winter I connect it all back to the sewer.

We think of using grey water as a classic case of recycling, but it's not that simple because grey water does add some unwanted chemicals to my soil. The day will come when each product that we buy is assessed for its usefulness further down the system. Ultimately our goal must be to move beyond recycling to 'source reduction'. Recycling is about using something again after I've used it once. Source reduction is not using something in the first place because, in the long run, you don't really need it, like most of the sodium in the laundry powder.

Playing around with waste water in the home is probably just for the very keen. I expect a lot of washing machine extension hoses will go to the back of the shed when the rain comes back and the dams fill a bit more. The greatest contribution to human health and wellbeing came from separating the clean water coming into the house from the dirty water leaving the house. There is no way a commercial domestic waste water system could pay its way in a home garden. So those of us dabbling with home-made grey water systems are blurring the margins.

Yet this dabbling is important because we can put pressure on laundry detergent manufacturers to make more environmentally-friendly products. For example we can use potassium instead of sodium in detergent. Obviously there is some impediment to making the conversion now, but I would pay a lot more for a potassium-based detergent than a sodium-based one. I bet it would be possible to make a laundry detergent that doubled as a fertiliser. At least by the time we have local suburb level waste water recycling systems, we will be wised up sufficiently to put pressure on those companies that fill their laundry detergent boxes with stuff we do not need.

Chapter 13

At the end of the river

The Murray River drains a vast basin of one million square kilometres into the Southern Ocean. It is fed by the Murrumbidgee River bringing water a thousand kilometres to the east, sporadically from the Darling River bringing water from a thousand kilometres to the north, and countless other smaller rivers and tributaries. Historically, 400 000 litres of water poured from the Murray River into the sea each second of every day.

Such a flow of water sounds impressive, but it is much less than rivers draining similar size basins in other parts of the world. The Zambezi, for example, carries to the sea in 20 days what the Murray does in an average year. The flow in the Murray is also highly variable from year to year, so average flows are misleading. The basin is huge, but it is also relatively dry and flat. Most of the rain is sucked up by trees, grasses and agricultural crops, long before it reaches a river.

A small amount, less than 5% of all the rain that falls in the basin, runs off the land surface and finds its way to a stream or river. Most of this run-off comes from the wetter and steeper south-eastern part of the basin, which makes up just 10% of the total area. In the days before dams and irrigation, just over half of the water entering the river system actually made it to the sea. The Murray is a slow moving meandering river, and nearly half of the water evaporated from wetlands, riverine vegetation and floodplains along the way.

When a big river runs through an arid land it seems obvious to scoop some water out and irrigate crops, before the water is lost to the sea. Great civilisations have sprung up in major river valleys around the world, but they have also faltered as their irrigation systems failed. The area known as the Fertile Crescent, or modern day Iraq, used to support vast areas of irrigated land between the Tigris and Euphrates rivers. Now it's mostly wasteland. In fact, the parts of the world that have been irrigated for the longest are mostly in decline.

Irrigated agriculture produces over one-third of the world's food. Yet some argue that up to a third of all the irrigated land in the world has become degraded, and is producing much lower crop yields than we expect. There is something fundamentally unsustainable about irrigation, unless it is practised very carefully.

For the last six years I worked with a group of grape farmers adjacent to Lake Alexandrina, who are among the last irrigators on the Murray–Darling river system. The huge lake is fed by the Murray River and empties into the sea. Once a month, a member of the community sends us all an email with a graph showing the change in the lake's water level.

On average, Lake Alexandrina is about 75 cm above sea level. As water evaporates from its huge surface, and as irrigators pump water for their crops during summer, the water level drops to around 50 cm above sea level. The winter rains in the Murray–Darling catchment re-fill the lake to around 90 cm above sea level.

In the summer of 2007 the lake level dropped much faster than usual because the inflow from the River Murray was so low. By autumn the lake was at sea level and it only recovered to 20 cm above sea level the following winter. During the summer of 2008 the lake dropped below sea level. Inflows from the Murray had virtually stopped and only the sand dunes and barrages across the river mouth prevented the sea from pouring in. The worst case scenario had unfolded, and few were prepared for it.

Our monthly email also showed us the salinity of the lake water. Before the drought, about 550 grams of salt was dissolved in every cubic metre of lake water. This sounds like a lot, but if you drank the water, it would barely taste salty. When the irrigators pump water from the lake for their vines, they add this dissolved salt to their soil. A single vine will happily use 3600 litres over a summer, and this much water would contain about 2 kg of salt. As the flow of water to the lake has dwindled in recent years, the salt content of the water has started to rise, and so has the salt added to the land.

Vines, like almost all plants, do not like salt. Salt attracts water to itself. We use salt as a preservative because the bacteria that rot our food cannot get the water they need from a salty brew. So vines in salty soil experience drought even when the soil is wet. Roots can still absorb some water and nutrients from a salty soil, but at the same time filter the salt out.

Salt is harvested for human consumption by flooding shallow bays with sea water and letting the water evaporate, leaving the salt behind. A similar process happens with irrigation. No water is completely free of salt. Each time the farmer irrigates the land, some salt is added to the soil, fresh water is transpired and the salt is left behind. A small fraction of the salt does slip into the plant and is carried to the leaves. Water evaporates from the leaves, leaving the salt trapped behind. Day after day a little more salt arrives in the transpiration stream until the leaves can take no more and die.

Meanwhile salt builds up in the soil season after season, unless we can provide extra water to flush it away. Leaching of the salt happens naturally in some places, particularly in winter rainfall areas. Since the vines do not use much water in winter, the rains quickly re-wet the soil. As the rain percolates through the soil, the accumulated salt dissolves and moves down with the water, hopefully beyond the root zone. If there is insufficient rain to leach the soil, irrigators can apply extra water, over and above what the plants actually use, to move salt out of the root zone.

But there is a catch. Where does the salt we are flushing out of the root zone end up? The soil may be one metre deep or 10 or even 100 metres deep, but eventually we reach an impermeable clay or rock. Then what?

Soil water is present as thin films around soil particles and within small pores between particles. The larger pores tend to stay filled with air. When the soil is wet, gravity pulls this water downwards until the impermeable layer is reached. Here, with nowhere else to go, the water fills all the large pores and displaces the air so the soil becomes saturated.

Water is 'wicked' sideways by dry soil away from the dripper. As the diameter of the wetting pattern increases, the downwards force of gravity becomes larger than the sideways wicking force. At first, the cross section of the wetting pattern is a semi-circle. Later it becomes much deeper than it is wide.

Saturated soil below the ground surface is called ground water, and the top of the saturated layer is called the water table. If, over a period of time, there is more rain or irrigation than the plants transpire, then water moves down towards the saturated layer. Eventually the water table starts to rise towards the soil surface.

Scientists who study water are called hydrologists, and there are three basic types of hydrologist. Surface hydrologists study the water that does not sink into the ground, but runs off the land surface. This water feeds rivers and also carries sediment. Surface hydrologists are excited by intense storms, because a lot of water runs off the land in a short time, eroding soil and swelling rivers.

Ground water hydrologists are only interested in water once it has entered the soil and percolated down to an impermeable layer. Here the water forms vast underground lakes, called aquifers. Compared to surface hydrology, things happen slowly in the ground water – many aquifers have taken thousands of years to fill. Springs and wetlands are formed on slopes or valley floors where aquifers meet the land surface. The aquifers leak water slowly into rivers, keeping them flowing through the long dry season.

Between the soil surface and the ground water is the domain of soil water hydrologists, who study the water held by soil that keeps plants and trees alive. Surface water, soil water and ground water all behave differently, and over very different time scales, and this is what makes the understanding of irrigation a bit complicated. Yet if we want to irrigate properly, we must have a basic understanding of all three.

We can get a feel for how surface water, soil water and ground water behave by digging a hole on a sloping beach a few metres above the water line. First we dig through moist sand, then wet sand, then sloppy, saturated sand. Soon the base of the hole fills with water. This marks the depth of the water table. If you pour a bucket of water out next to the hole, a little water will run on the surface down-slope towards the sea, but most will seep vertically downwards into the sand.

If you pour water directly into the water at the bottom of the hole, water will seep away sideways down towards the sea. In other words, surface water and ground water tend to move down a slope as we would expect of any liquid. Soil water is different. It is attracted to soil particles, and is pulled vertically to the water table under the force of gravity, regardless of the slope of the land.

When we added water directly to the hole on the beach, the water level rose, and then fell back to its original level quite quickly, as the water seeped sideways towards the sea. If we did the same experiment by digging a hole on a sloping clay soil, the result would be very different. When we pour water on the surface, a lot more of the water would run off down the slope rather than infiltrate into the soil. And when we add a bucket of water directly to the hole in clay soil, the hole would fill up with water and could well be full of water for the next few hours or even days.

Sand contains large pores that allow ground water to seep sideways quite quickly, but clay has tiny pores, and water moves very slowly through it. So ground water in a coarse material will behave very differently to ground water in a fine clay material. Ground water in sloping country also behaves differently from ground water in flat country. This has far-reaching implications for irrigators.

High water tables are common in irrigated areas and the ground water can be salty as well. Perforated pipes are buried in the soil so that the ground water can be collected and pumped away from the river into giant evaporation ponds.

This difference between fast and slow, downwards and sidewards may sound a bit trivial, but it has brought down civilisations when they got it wrong. Historically, most irrigation systems fail when they eventually drown as ground water rises around them, not because they run out of water. Irrigators tend to apply more water than the plants need, and this extra water moves down to the ground water faster than the ground water can move sideways to a river.

The balance between water moving downward towards the aquifer and sideways towards a low point in the landscape can take decades to play itself out. When more water arrives at the water table than can be moved sideways through the aquifer, the water table will rise. It may start 30 m below the soil surface, and rise at half a metre a year. For 50 years nobody really notices. Then the water table rises to the bottom of the root zone. Now, instead of irrigation washing salty water downwards, the salty water is coming up.

This happens all too often. In large flat basins like the Murray and many others around the world, sideways movement of ground water is very slow indeed. From the perspective of the irrigator, applying too little water means low yields. To stay in business, it's better to err on the side of too much water than too little. In the shorter term the extra water just seeps away. In the long term, however, over-irrigation can be a disaster. Sometimes the land gets water-logged. More often it's a deadly combination of salt and waterlogging.

There are important food producing areas, particularly in India and China, where the ground water is not salty, and here farmers face the problem in reverse. Ground water presents an irresistible resource for irrigators, who sink wells to pump water for their crops. This would be all right if the extraction of water from wells matched the replenishment of the ground water from rain percolating below the root zone of crops. Rarely does this occur. As fresh ground water levels fall, pumping costs escalate, and the wealthier irrigators just sink deeper wells. It is thought that 10% of the world's grain is produced from water that filled aquifers decades or centuries ago, and will never be replaced.

Whether we apply water from rivers or ground water, irrigation is ultimately a salt-concentrating business. We apply slightly salty water to the crops. The crops use the water and leave the salt behind. The origin of the salt may be from weathered rocks, or from soils once covered by sea, but most of the salt in the Murray Basin has arrived in the rain. Small amounts of salt, whipped up by storms over the oceans, are carried over the continent by rain-bearing winds. Since almost all the rain is transpired by plants, the tiny amounts of salt build up over thousands of years. Long before irrigation started in the basin, early European explorers found parts of the river too salty to drink in times of drought.

Irrigation contributes to the saltiness of the river and those at the end of the river system have to put up with all the problems generated from people upstream. The grape growers around Lake Alexandrina were well aware of this.

One of my students is a Mozambican working with the last irrigators on the Limpopo River, and grappling with the all too familiar salt and waterlogging problems. Looking back upstream, he faces a more bewildering set of circumstances than those facing the Murray–Darling Basin. The Limpopo and its tributaries drain parts of Botswana, Zimbabwe and South Africa, before crossing into Mozambique along the northern boundary of Kruger National Park.

In 2007, crocodiles started dying in large numbers in one of the major tributaries of the Limpopo. Some point the finger at toxic heavy metals leaching from the mines or agricultural chemicals from the irrigation areas. Others are concerned about the increasing failure of water treatment plants, spewing raw sewage into the river. There is a theory that the dams are implicated. Large numbers of bottom-feeding fish moving upstream as dams fill are eaten by the crocodiles, and this seems to set off the problem.

The crocodiles that die are adults and almost always very fat, so it does not seem as if they are sick. The vets can describe the pathology of the disease and have a name for it, but have no idea what causes it or how to stop it. The crocodile is iconic, not because it is cute and cuddly, but because it has until now been seen as indestructible. It is a real worry when the collective management of the river is killing the top predator and we have little clue as to why. The story has an ominous presence as Australia looks to develop its northern rivers.

It is a major challenge to share the water between irrigation, mining, industry, urban, forestry, and conservation interests across four nations. Most years, large reaches of the Limpopo River actually stop flowing, and drought is debilitating to subsistence farmers. In 2000, half a million Mozambicans lost their homes and livelihoods when the Limpopo flooded. Development tries to reduce these risks by reducing variability, by building dams and regulating flow.

Society gets many benefits from mitigating floods and droughts, but rivers are biodiversity hotspots, and it is variability – the droughts and the floods – that maintain habitat variability and hence the diversity we value. And invariably, more dams means that more water will be used for irrigation, which exacerbates the problem in the long run.

The days of building new dams in southern Australia are gone. Already the existing dams can hold three times the total annual river flow in the Murray system, and there is not enough water to fill them. Moreover the flow in the river is almost certain to decline. The biggest impact is expected to come from climate change. Plantation forestry and vigorous bush regrowth after the 2002–03 bushfires and the impact of small farm dams will reduce the flow of water to rivers. Irrigators are tapping into the better quality ground water which used to feed the river, and more efficient irrigation systems ensure that more water is transpired by crops and less finds its way back to ground water and the river.

In an average year, 46% of the water in the Murray–Darling Basin is used for irrigation, 6% evaporates from storages and 28% evaporates from wetlands. Once, 54% of the water reached the sea – now the average is 20% and falling. But these days no one knows what average means. For two summers in a row almost no water has reached Lake Alexandrina, brought about by an unknown combination of the current drought and climate change. Best estimates are that we need to take at least 25 to 50% less water out of the river if we want to save both the irrigation industry and the river that sustains it.

From time to time there is a clamour to divert water from so-called well-watered areas to those wracked with drought. The Snowy Mountain Scheme diverts water into the Murray–Darling Basin that would have flowed east to the sea, augmenting supplies by about 5%. The proposed schemes to redirect the flow of northern rivers would cost so much that we could barely afford to drink the water they would provide, and the ecological cost to the catchments losing the water is always underestimated.

Yet it is not the cost that makes these schemes so outlandish, but the failure to learn the lessons of the past. The landscape may look parched and dry in a drought, but it's more likely that we have added more water and salt to an irrigated landscape than ground water can carry away. For many irrigation areas, the problem in the long run is too much water, and the attendant problems of waterlogging and salinity. In most parts of the world, the future of irrigation will have to be making better use of less water.

Chapter 14

The Goldilocks principle

If my neighbour keeps too many cows, and they eat the grass faster than it grows, he will soon run out of grass. If I keep fewer cows, the pasture on my side of the fence will thrive while his side will be denuded. But if my neighbour irrigates his crops with more water than they need, then the water table may rise. It will not stop at the fence line, but rise under my farm and his.

For this reason, we have to think of water at whole-of-catchment scale, and irrigators in the same district need to work together to safeguard their resource. This is exactly what the grape growers around Lake Alexandrina decided to do. They decided to set up an Irrigator Code of Practice, so that they could have some control over how water was used in their local area. Every irrigator in the district signed up.

They already knew the four rules of sustainable irrigation. First, where the ground water is fresh, don't pump it up faster than the rain can replenish the aquifer. Second, salt will accumulate in the soil in a semi-arid environment and extra water is needed to flush the salt out of the root zone. Third, the salt eventually ends up in the ground water. Fourth, when water arrives at the salty ground water faster than it seeps sideways to the river, the water table and salt will rise.

The grape growers around Lake Alexandrina had already faced down problem one. Their ground water was relatively fresh in some areas and was salty in others. They used to pump water from the relatively fresh aquifer, but the levels were falling and the ground water was getting saltier.

So they did a deal. They swapped their entitlement to use ground water for a licence to use river water from the lake. It cost a lot of money to put in all the pipes to make this conversion, but it was preferable to depleting the fresh aquifer some time down the track.

Now they had to prepare for problems two, three and four. Like all irrigators they would be adding salt to their vineyards with the irrigation water, and the lake water contained a reasonable concentration of salt. This salt may accumulate in the root zone of the grapes. If too much lake water was used, the salty ground water levels could start to rise.

They set up a Water Management Committee made up of local growers, which would develop and administer the local Irrigator Code of Practice. The first requirement was for each irrigator to disclose the total amount of water they used on their vineyards in a year. At the end of the season, each irrigator received a graph showing the amount of water used per hectare by every other irrigator in the district. They could identify themselves on the graph, but no one else. The idea was that each irrigator could anonymously see where they slotted in among their peers.

The second step was for each irrigator to sink two wells on their property, six metres deep. Four times a year, they had to drop a tape measure into the well, measure the depth to the ground water, and report this to their local committee. Ground water hydrologists would not have been too impressed by this. There are strategic locations to place wells based on geological knowledge, not the whim of the farmer. Moreover six metres was far too shallow for most parts of the district. You need bores 20 m or 50 m deep to really know what's going on.

For the committee though, this was an experiment in social learning as much as an exercise in hydrology. The farmers had to fund the sinking of their own wells and the cost limited the depth to which they could go. They had heard all kinds of scare stories about water tables rising here and there, but it was all out of sight and hard to fathom. By measuring for themselves, the slow and strange behaviour of ground water became tangible. The information they got from their wells did not always make sense, and two wells close by could give different answers, but through experience they began to unravel some of the mysteries of ground water. When the experts came to talk, they knew the right questions to ask.

The third step was an agreement to plant a certain area of native trees in proportion to the amount of water they were using for irrigation. The trees could be planted on their own farms, or they could provide the funds to revegetate a specially selected area of communal land in the district. This was in recognition that irrigation has far-reaching implications, well beyond the boundaries of their own farms, and that trees could play a role in stopping ground water from rising.

Around this time, the committee got hold of an early prototype of a wetting front detector, a few years before it became commercially available. One of the locals disassembled the detector and then constructed 20 copies. These were buried around the district. Six months later I was astounded to receive a letter from the Water Management Committee requesting 300 more detectors, needing them immediately, and not needing any guidance as to how to use them. Apparently they already knew.

The committee had decided to expand their Irrigator Code of Practice. They already recorded the annual water use, plus depth to ground water from each property. They had implemented the requirement to plant native vegetation. Now they wanted to get a feel for whether salt was accumulating in the root zone.

Each grower had to purchase two wetting front detectors and a local grower did the installation across the district. Then all the grape growers were asked to report the date of every irrigation event and how long they irrigated for. After irrigation, they had to report whether a shallow detector buried at 50 cm depth or a deep detector buried at 100 cm depth collected water. If the detector at 100 cm depth recorded a wetting front, they had to suck the water out of the detector and measure the salt content. If they did not have a salt meter they could drop the sample at the local post office, and a member of the community did the measurement for them.

This was exciting stuff. We scientists normally spend hundreds of thousands of dollars of research funds to organise the collection of such data across a whole district. This one had sprung up spontaneously from the grassroots. A year into the program I was invited to speak at the annual general meeting for grape irrigators in the district. About a hundred farmers turned up to the local football club.

Just before it was my turn to speak, the MC asked which irrigators were satisfied with the wetting front detector project. A sprinkling of hands went up around the audience. I thought the farmers were a bit shy. After all, if you put your hand up in a public meeting, you may be asked a question which you subsequently can't answer.

Then the MC asked who were not satisfied, and half the hands in the room went up. This was the first test of the detector on a large scale, albeit in prototype form, and at least half the farmers were unimpressed. It turned out that about 20% of participants had not captured any water in their detector at 50 cm depth, and substantially more had not seen any action at 100 cm depth. The project required considerable effort from the farmers, but was apparently delivering little in return.

I gave my presentation on measuring and managing salt in the root zone. Then, at the end of the meeting, the MC asked those who were really dissatisfied to write their name down and announced that I would personally visit them.

I had a troubled night, remembering Einstein's admonition, 'make it as simple as possible but no simpler'. I had spent a long time trying to develop the simplest tool that would make water and salt monitoring come alive for the majority. The farmers saw it and immediately liked it. But was it too simple? Would it stand up in the real world?

The principle of a buried funnel with a sealed reservoir at the base seems simple enough. You would think that water moving downwards through the soil would eventually get caught by the funnel. Once captured by the funnel, the water would fill the reservoir and activate the float; however, a funnel buried in the soil behaves very differently from a funnel in the open air. If it is drizzling with rain, a funnel with a cork in the bottom in the open can only collect water. It may collect fast or slow, but once the water enters the wide end of the funnel it is trapped.

It is not the same when a funnel is buried in soil. Water percolates downwards as thin films around soil particles. These films of water are converged by the funnel, so the soil inside the funnel gets wetter as the funnel narrows, in much the same way as a beam of light is focused by a magnifying glass. But soil is also able to wick water upwards and back out of the funnel. If we suspend a piece of cloth in a glass of water the water moves upwards from wet to dry.

So we have two opposing forces at work. Convergence makes the soil in the funnel wetter and wicking tries to get some of the collected water back out. If the convergence cannot fill the funnel faster than the wicking soil can empty it, then the detector records nothing.

The deeper we place a detector in the soil, the harder it becomes to capture a water sample. This is because the wetting front gets progressively weaker with time. You can visualise this by taking a bone-dry, stiff facecloth and dipping one edge into water to a depth of 5 cm. Then suspend the cloth vertically, as though pegging it on the line to dry, with the 5 cm wet strip at the top. At first there will be a sharp line demarcating the darker coloured wet cloth above from the lighter coloured dry cloth below. This line marks the wetting front.

As water is pulled downwards by a combination of the wicking power of dry cloth and gravity, the wetting front becomes weaker; the line dividing wet from dry cloth first becomes diffuse and eventually hard to see at all. Clearly, many of the detectors had been buried too deep and I was about to face dissatisfied farmers and explain the error.

The first four visits all turned up plausible reasons to explain why the detectors were not collecting. Drip emitters were not placed right above the detector, or were applying much less water than expected, or the water was running across the surface and infiltrating some distance from the detector. But it was the next two visits that really opened my eyes. Farmer five declared we were lucky, because it had rained heavily a few nights before and the detector had popped up its float for the first time ever. I removed the sample of water and tasted a drop. It was very salty. The story at farmer six was the same, so I took a sample there too.

At first we were puzzled when some of the wetting front detectors collected no water. Then, when we did get water samples, we found they contained lots of salt. Irrigators who applied water little and often found salt building up in the root zone.

Back in the lab, I tested the salt content of the water, and for both farms it was nearly half the salinity of sea water. Normally we would be worried if we measured one-tenth the salinity of sea water. This high salt reading was a tell-tale clue. Each time the irrigation is turned on, the wetting front picks up any salt in the soil and carries it forward, like a snow plough shovelling snow. As the wetting front dissipates after the irrigation is turned off, the salt is deposited near the edge of the wetting pattern, like driftwood on the beach. A little more salt is added with each irrigation and shunted towards the perimeter of the wetting pattern.

The only way we could have collected such a salty sample from these farms was if salt had been accumulating above the 50 cm depth detector for some time. And if salt was accumulating above the detector, then not much water would have been going past the detector.

We had, no doubt, put the detectors too deep, and were collecting fewer samples than we should, but I felt I had some understanding of what was happening. The Lake Alexandrina growers prided themselves on using a small amount of water and thus being efficient irrigators. Keeping the grapes under some water stress increased the quality of the resultant wine. But perhaps they were too efficient. Applying less water than the vines needed coupled with a few dry winters meant there was not much leaching and, in places, dangerous quantities of salt were building up in the root zone.

There are three ingredients to learning-by-doing. First we have to develop a shared view of how we think the system 'works'. Then we have to monitor the right things so we can test our ideas. As we share experiences, we see how well our understanding fits with reality. It's the surprises that show there is much more to learn.

Over the following six years the grape growers provided us with more and more salt readings. The information was proving useful and a general pattern was emerging. Those who activated their shallow detectors the least, also measured the most salt when they finally did get a reading. It would make sense for these farmers to apply more water less often, to push the salt a bit deeper. And the minority who activated their deep detectors on a regular basis tended to have the lowest salt. These farmers could apply less water at one time.

As the amount of water reaching Lake Alexandrina has dwindled to a trickle in recent years, the salinity of the lake water has shot up. The water is quickly becoming too salty to use. When a river starts to die, this is seen first at the bottom end. No water out the mouth means there is no dilution of the salt, which now accumulates in irrigation regions and along the floodplains of the river.

It is somewhat ironic that the Lake Alexandrina grape growers, with their Irrigator Code of Practice, should pay a heavy price for the development obsession that existed in the Murray–Darling Basin between the 1960s and the 1990s. Successive State governments handed out water licences in isolation, with little idea that the sum total of them exceeded the water needed for irrigators and the health of the river.

There have been a few lessons for me as well. The detector seemed to be failing, at least in the beginning when few samples were collected. I don't like failure, particularly in public, but it tells you what you most need to know. If you can live with it for a while, failure helps you to learn a little faster, and the experience taught me much about how to interpret the data we were getting. It spurred us on to define the sensitivity of the detector, specify appropriate placement depths, and even how to build different types of detectors for different situations.

Yet there was an even greater lesson. The build up and dissipation of salt in the root zone was giving us an insight into the efficiency of irrigation management. Salt should build up in the root zone, at least in the lower part. If it does not, then too much water is moving below the roots. This water is wasted, may contribute to rising ground water and is probably carrying nutrients as well. Only when the salt has built up to a threshold value should we flush it a bit deeper into the soil. Monitoring salt can help us to become much better irrigators.

This realisation gave rise to what I call the Goldilocks principle. The Goldilocks principle means you must keep track of three types of variables that change over different time scales. The fast variable is linked to productivity (grapes) and the slow variable is linked to sustainability (salt and ground water). Then there is the Goldilocks variable between – not too fast and not too slow – something that helps us learn, giving us the understanding that will help us to alter course.

I started my research career trying to get irrigators to focus on soil water monitoring. This is a fast variable – the soil water content goes high and low within a week between irrigation events. Keeping this variable under control – the right amount of water in the root zone of an irrigated crop – is essential for getting good yields and staying in business.

Then my work swung over to the slow variable. I started work on how agriculture affected the gradual rise of ground water and the movement of salt into root zones and rivers. This variable typically plays itself out over decades or more and ultimately sets the limits to sustainable production.

The trouble with the fast variable is that it is too fast. We make decisions to maximise productivity without realising we may be destroying the very ecosystem that sustains us. The problem with the slow variable is that it is too slow; too hard to make expensive decisions now that may only pay a dividend for the following generations or perhaps no dividend at all.

Like Goldilocks who found the porridge too hot or too cold and the bed too hard or too soft, I was looking for a variable that did not change too fast or too slow. It turns out to be salt concentration near the bottom of the root zone. Salt rises and falls over a season or several seasons, slower than changes in soil water content but much faster than rises in ground water level.

If salt never builds up near the bottom of the root zone, then too much water is going through my soil, adding unnecessarily to water tables and probably wasting nutrients too. If salt builds up too much, I have to leach it, and understand where the leached water will be in one or 10 year's time. This is the scale variable that, when coupled with the fast and slow one on either side, helps us to understand how the whole system behaves.

PART 3
FEEDING OURSELVES

In my line of research work there has been tension between those who focus on the productivity of agriculture, and those who look at the ecological footprint of agriculture. During the years when food seemed plentiful, and we were uncovering the extent to which agriculture has affected the environment, the environmental side of the debate had the upper hand. Then came the global food crisis and there were food riots in over 30 countries as the price of rice, maize and wheat soared out of the reach of the poor.

This collection of essays delves into what we eat and why, and who produces our food. Agriculture means we cut down trees to make way for the plough, so we explore tillage, permaculture and agroforestry as well. I have spent a lot of my time working in Southern Africa, so we also look at agriculture through the eyes of the resource poor farmer.

Finally we deal with making sense of an increasing complex world, before returning to the garden for a short lesson on growing more food with less water.

A lot of food was once grown close to where people lived. Mixed farming, the combination of animals and a wide variety of crops, was common because manures were needed for plants and variety reduced risk. The industrial food chain increasingly provided economies of scale, so farms tended to specialise, so that whole regions now grow just a few crops. Diversified farming systems, such as permaculture and agroforestry, are yet to make a major contribution to global food supply.

As farms get bigger, people are leaving rural areas for the cities. The large farms are increasingly reliant on bigger machines and external inputs of chemical fertiliser and fossil fuel energy. Meanwhile the rural poor eke out a living at the margins. They lift water from shallow wells, use foot-operated pumps to water small plots, scoop water from the irrigation canal to wash their clothes and fill containers of drinking water loaded on donkey-drawn carts.

A generation ago, most people in the western world had some direct connection with how their food was produced. They may have come from a farm, had relatives on the land or at least spent holidays in the countryside. The food on our plates these days is likely to have been sourced hundreds or thousands of kilometres from where we live. There is an upsurge of interest in what we eat, but few know how our food is grown and processed.

My own journey through this book starts and ends in a garden. I wrote from where I could see the garden, held meetings in the garden and shared meals in the garden. The garden is a sanctuary, shielded from the busy world outside – a productive place, a place to play, to experiment and to learn.

Chapter 15

A short history of agriculture

In the history of agriculture, 2007 was a watershed year. Before then, more people lived in rural areas than in towns and cities. Today, the urbanites are in the majority and the rush to the city continues apace. Better crop varieties, farming methods, tools, agrochemicals and mechanisation have, over the years, allowed each farmer to produce more food, on larger farms, with fewer labourers.

The bright lights of the city have proved irresistible for the erstwhile farmers, the surplus farm workers and the landless rural poor. Many now labour in factories instead of fields, producing an endless stream of manufactured goods to make our lives easier or more enjoyable. The drift from country to city is reaching its limit in developed countries, where fewer than five out of 100 workers are involved in food production. The majority of the population in many developing countries is still involved in farming, but the move to the city is relentless.

Based on archaeological evidence, agriculture got going around 10 000 years ago. Prior to the development of agriculture, people lived by gathering wild food and hunting animals. Hunter–gatherers probably lived in small kinship groups and followed a nomadic lifestyle, searching for edible plants and following animals. There is not much point in having a lot of possessions if you have to carry them, so hunter–gatherers lived a rather simple life.

The vast majority of plants are not edible, so hunter–gatherers foraged over large areas, and the population density was necessarily low. The development of agriculture involved collecting the few edible plants and trying to grow them in one place. Surrounding the home with edible plants would save an enormous amount of time searching for them in the forest or savannah.

By collecting and then sowing edible plants around their temporary dwellings, the hunter–gatherers inadvertently selected for attributes that suited them and against those that did not. For example, the ancestors of the wheat plant had smaller seeds than modern wheat varieties. Moreover the seeds fell out of the head as soon as they were ripe, so they were difficult to collect. Many of the seeds remained dormant in the soil, which means they did not immediately sprout when it rained.

Attributes like dormancy and a large number of small seeds that were easily dislodged and scattered by animals, helped the wild wheat plants to survive. More seeds scattered more widely increased the chance of a few finding a good spot to grow. Moreover, if every wheat seed germinated at the first shower of rain which was then followed by a long dry period, then all the little seedlings would die. But when the seeds have a range of dormancies, then different seeds will germinate weeks, months or sometimes years after they fell from the plant. Some seeds at least are bound to germinate during a period when the rains are reliable, and will grow to maturity, setting seed for the next generation.

The early farmers were more likely to collect seeds that stayed in the head when ripe, rather than falling to the ground. They planted the largest seeds at the start of the wet season, and tended the plants that germinated first. So these farmers selected for traits that were beneficial to them, and selected against some of the traits needed for survival in the wild. Generation by generation, the plants evolved into the modern forms we know now, a process we call domestication.

The earliest site for the start of agriculture was probably South-East Asia, or the area called the Fertile Crescent. The wild ancestors of wheat and peas, and the ancestor of modern sheep and goats, were native to this region. A thousand years later there is evidence of rice being domesticated in China, and soon after, the domestication of soybeans and pigs. In Central America corn, beans and turkeys were domesticated, and in South America potatoes, cassava and guinea pigs.

The common theme in the above is that there had to be a few native plants available in each region which could be domesticated before agriculture was worthwhile. It would not be possible to give up the hunter–gathering life if you just had some primitive wheat growing in a field nearby. Ideally you needed a starch crop (wheat, rice, maize, potato), a legume for protein (pea, soybean, green bean) and an animal. The combination of all three was rare in one place, so the development of agriculture was rare too.

Rice and wheat produce most of the world's food energy. Selective breeding to produce shorter crops that can hold larger heads started the Green Revolution. These new varieties are more responsive to fertilisers.

The most important staple food crops are rice, wheat, maize, oats and sorghum and these are all part of the grass family. These grassy species are annuals, and are mostly adapted to places that have a wet and dry season or a warm and cool season. In places that are too cold or dry for crops year round, farming was based around animals. Animals can forage for themselves and walk long distances to gather the feed they need from inhospitable places. During short periods of abundance they put on weight and have their young, which are then culled throughout the year to even out the food supply. Animals become the buffer between the herders and their harsh environment.

Some societies never turned to cropping or herding, but remained hunter–gatherers. One example is the Aboriginal Australians, who collected wild roots, tubers, grass seeds and acacia pods, and hunted kangaroos right up to modern times, using Stone Age tools. At first sight this seems to show a lack of initiative, but that is not true. With all the help of modern science, no-one has domesticated any Australian plant or animal, except the macadamia nut. In the absence of native species that could be domesticated for agriculture, and separated from the Asian mainland from which they could have sourced them, the Aborigines became experts in understanding the land. To survive in a dry and infertile land they had to be expert naturalists.

Once plants were domesticated, they probably spread quite quickly, at least within continents. The more plants and animals available, the more successful agriculture became, and population density increased. Unlike the hunter–gatherer clans, where almost everyone was involved in foraging for food, not everyone had to farm full time. They could be part-time artisans or part-time soldiers and there was usually a ruling elite that did not farm at all.

With their greater manpower and organisation, agricultural societies tended to displace or absorb hunter–gatherer societies. There were a few instances where hunter–gatherers had the advantage. South American Indians survived in the Amazon which was too wet for agriculture, Bushmen in the Kalahari where it was too dry and Inuit in the Arctic where it was too cold. Frequently it was diseases like malaria that kept the farmers out, or cattle diseases like sleeping sickness that kept the herders out.

Though agriculture proved more successful than hunter–gathering, it was still very hard work. Those who had the chance preferred to get out of agriculture all together. For example, the Roman armies captured many slaves, and put them to work in the fields. A class of wealthy land owners emerged, who assumed roles in the urban political scene and bureaucracy. The centre of power and influence quickly shifted to the cities.

In the 16th century, the world had to feed half a billion people. By the 19th century there were over one billion mouths to feed and by the end of the 20th century six billion. The rate of population growth is slowing, but the number of people on earth still goes up by 200 000 mouths per day. So agriculture has to get more and more productive, and by and large it has met the challenge. There has been starvation on a large scale, but the problem is usually that a poor cropping season hits a country already crippled by war or unrest, and tips it over the edge. There are large numbers of under-nourished people in the world, but this is usually an economic, not a scarcity problem.

Agriculture may have been successful in feeding the world, but it has come at a cost. First, the area of agricultural land has expanded dramatically. It has been estimated that 40% of all the photosynthesis that occurs on the planet is diverted to feed people, which of course has a major impact on all the other species on the planet. More than half the world's accessible fresh water has been diverted to agriculture. In many places there is no more land to farm and no more water to pump. Agriculture simply cannot keep expanding.

Land is also degraded by agriculture. Ploughed and overgrazed soils are eroded by wind and water. Soils are compacted by heavy machinery and the hooves of domestic stock. Soil nutrient levels fall as the harvested food is exported to the city and the city's sewage dumped into the sea. Many soils that have been cropped and fertilised for a long time start to go acidic. Irrigated soils tend to go salty or get waterlogged.

A few years back, a brave attempt was made to put a cost on all this for Australia. If, for example, we could just wipe away the problems of acidity and salinity, then our farmers would make an extra $1.6 billion a year through higher yields. Degradation does not only affect the farm itself, but has downstream costs as well. The cost of declining water quality to water suppliers and irrigators through increasing salinity, turbidity and sedimentation could soon be costing a further $2 billion per year.

When we think that the net profit from agriculture is about $6.6 billion per year, then the cost of degradation eats deeply into the cake. Add in a further $2.2 billion of assistance for the agricultural sector in the form of tariffs, government support, extension and research and the profit from the agriculture sector looks thin indeed. And this is before we factor in the things we love but find hard to value, like wildlife, rivers to swim in and fish, aesthetically pleasing landscapes and the like, which are threatened by the expansion of agriculture.

Degradation is often communicated in emotive terms as 'football fields per hour' of land lost to salinity, or erosion measured as 'kilometres of soil-laden semi-trailers parked end to end'. Put like this, agriculture sounds like a disaster, and the task of feeding a growing population hopeless. The reality is that degradation is very serious, but is somewhat counterbalanced by innovation, as the following example shows.

When farmers first started growing wheat in Australia, they were farming virgin country. Yields were not high by world standards; the grain from each square metre of crop would be sufficient for about four large slices of bread. Each decade the yields declined, mainly through exhaustion of the nutrients in the soil, so by 1900 just two slices of bread came from each square metre of crop.

Then the trend started to reverse, led by plant breeders who developed cultivars of wheat that were better matched to the Australian environment. Farmers also learned to fallow the land, and thereby allow the soil water reserves to build up between crops. Then soluble phosphorus fertiliser became available for the first time. By 1950, the yields had almost reached the original four slices of bread per square metre.

After this, yields started to rise even faster. One of the main reasons was the development of clover-based pastures in rotation with wheat, which greatly boosted the nitrogen status of the soil. Bigger tractors and harvesters meant farmers could plant and harvest crops at the best possible time. Herbicides became available for the first time.

In the 1960s there were further improvements in the wheat cultivars and new crops were introduced, like lupins, which could be grown in rotation with wheat. By now the wheat crops were bigger than ever, yielding six slices of bread per square metre.

A new oilseed crop, canola, was developed from rapeseed in the 1980s, and became a profitable crop in the wheat belt. Wheat crops often grew better after canola than just about any other crop. Farmers expected wheat to grow better after clover or lupins because the benefits of legumes were well known, but canola was not a legume. It turned out that canola

A modern pea harvester can pick and shell five tonnes of peas an hour. The processing factory will have them in 5000 little frozen bags by the end of the day. Meanwhile, many subsistence farmers spend most of the day producing enough food to satisfy their daily needs.

inhibited the growth of some soil bacteria and fungi that had been nibbling away unseen on wheat roots. The canola acted as a mild form of soil fumigant, by preventing the diseases of wheat building up year on year. With the soil diseases under control, farmers found that their crops started to respond better to nitrogen fertiliser. By the end of the century average yields were eight slices per square metre, double what they had been on virgin land. Some farmers get more than double this.

We can look at the same example through a different lens. When European settlers first arrived in Australia it was a battle just to grow enough food to survive. Gradually, as the early breakthroughs kicked in, the country moved out of the survival phase and entered an era of productivity. Australia could sell all the wheat and wool it could grow.

From productivity the focus shifted to profitability. It was one thing to grow the crop, but in a competitive world market, could you actually sell it at a profit? Breakthroughs continued and yields went up, but not the prices, so many farmers were forced off the land.

By the 1980s the word 'sustainable' had entered common usage. Agriculture had gone through the survival, productivity and profitability phases, but the questions had shifted to the other side. Was agriculture sustainable? In other words, could we go on farming the same way without destroying the very soil and water resources upon which agriculture depended?

Sustainability was always a vague term. The well-known report, 'Our Common Future', called for 'sustainable development' so this generation could meet its own needs without compromising the ability of the next generation to meet theirs. But the world 'development' has a life of its own, depending on your viewpoint. We lose several kilograms of soil for every kilogram of wheat produced, so a soil scientist would say that is unsustainable. Yet the farmers must adopt practices that deliver certain yields otherwise the farm itself cannot remain sustainable. And as the farms get bigger and more profitable, the rural town gets smaller, and the local bank branch or high school becomes unsustainable.

More recently a new term to describe the relationship between food production and the wider environment has come into use: resilience. According to this view, we do not expect things to be sustained. In fact we expect change. Resilience challenges the idea that if we become very good at doing something, it will eventually be sustainable.

Irrigation is a good example. We may get so efficient at producing an irrigated crop that the whole district becomes involved. Over the decades, some water inevitably escapes to the ground water, no matter how efficient we learn to be. At first this is not a problem, until the water table has risen to two metres from the surface. Then just one or two wet years can tip the balance. The water table rises suddenly into the root zone, bringing the salt with it. Or conversely there are a series of dry years. The water resources are stretched to the limit; quotas are to be cut in half. In both cases, farms that were once considered efficient are in peril.

The point is that if we focus on doing one thing efficiently and sustainably and do more and more of it, we actually make ourselves more vulnerable to the inevitable shocks that are beyond our control. We need to invest in diversity, and foster the things that are not the most profitable today, because we will need them tomorrow.

Meanwhile, the realities of everyday economics dictate that people will continue to leave the profession of farming. This makes sense in poor countries where half or more of the population are involved in food production. Farmers will be poor because there are not many buyers for their surpluses. And if the majority of the population is living a subsistence existence there will not be much of a market for all the goods of the city.

It is not surprising that farmers leave the land. Until recently, a farmer was paid the same amount for wheat or rice as they received 15 years ago. Over the same period, all the other goods and services they need increased in price five times. So a farmer producing the same amount of food year by year will quite quickly go broke. They have to run to stay in the same place.

Farming is rarely as lucrative as a city job. Most farmers will tell you that it's a lifestyle or a vocation, and that keeps them on the land. In poor countries it's usually because there is no other choice. After all, the average income of the rural population is only one-tenth that of global average. So far, apart from localised famines exacerbated by war, farmers have grown enough food to feed the world. But it is not at all clear that agriculture has become more resilient to the inevitable, but still uncertain shocks of tomorrow.

Chapter 16

What we eat

Visitors who see our garden in mid-summer, when the fruit trees are laden and the beds overflowing with vegetables, have two immediate responses. 'Are you self-sufficient?' and 'How much time does it take to grow this much food?'

We are not nearly self sufficient, but I do spend a lot of time in the garden. For half the year we grow more fruit and vegetables than we can eat and half the year not quite enough. Although we do some freezing and drying, storing food requires almost as much art and time as actually growing it. In the past we have kept a couple of milk goats and eaten a few of the offspring, and we still keep laying hens. But the question remains: what would it take to become self-sufficient?

The first step is to produce enough energy to keep ourselves going. Humans are warm-blooded and it takes a lot of fuel to keep the body at 37°C. Just sitting around doing nothing uses energy at the rate of a 75 Watt light globe. On average, about 70% of all the food energy we consume in a day is needed to just keep the body organs ticking over. Another 10% is used up in digesting the food we eat. The remaining 20% of the energy we get from food is burnt up in activity and exercise. Very active people burn more, but the difference is not as large as you might expect. If you went from a sedentary office job to extreme athletics you would only need to eat 60% more calories.

Food energy is measured in kilojoules or kilocalories (there are 4.2 kilojoules in a kilocalorie; I use the shorthand Calories – big 'C' Cal – for kilocalorie). A grown man needs around 2500 Cal each day to remain the same weight. That's enough energy to boil 30 litres of water from room temperature, assuming no heat loss through the sides of the container. If we want to be more accurate, there are equations to calculate the number of Calories we should eat each day based on age, sex, height, weight and level of activity. When I run my family through the equations it comes to around 11 500 Cal per day or 4.2 million Calories in a year.

Our property is 877 square metres, a bit smaller than the proverbial 'quarter acre block'. The house, driveway, deck and surrounding hard surfaces occupy about 327 square metres, leaving 550 square metres for agriculture, assuming we plant every available space. The most energy packed vegetable I could grow in my climate is the potato. An acceptable yield would be about 3 kg of potatoes for each square metre per year and at 17% carbohydrate and 2% protein, is worth around 760 Calories per kg. The total calorific output from my fence to fence potato garden would be around 1.2 million Calories for the year. That might keep two of us alive, although we would become ill for lack of protein and other essential nutrients, not to mention the boredom of eating 3 kg of potatoes each day.

If we switched to wheat production, we could grow almost as many Calories as potatoes and substantially more protein. At a push we could grow a wheat crop through the winter and immediately thereafter sow potatoes which can be harvested just before wheat planting time. The garden is now under constant cultivation, requiring inputs of water, fertiliser and pesticides, and could provide about half the food energy needed to keep the family alive and active.

It is, of course, impossible to sustain a family of five on one-twentieth of a hectare of suburban garden, no matter how fertile the land is. In Australia there are 40 hectares of land per person, if we divide the total area by the population. Much of this land is too dry to be farmed, but there are still around 22 hectares that are grazed and just over one hectare under crops for each Australian. Australia is a country where more food is exported than locally consumed, but the global statistics are more sobering.

An average football field is 100 m long and 50 m wide. Two football fields side by side make a 100 m by 100 m square, or exactly one hectare. Twenty-five years ago there was just over one hectare for each of us, but since the early 1980s the world has added another 2.2 billion people. Now there is three-quarters of one hectare of farmland each; one football field and from the goal line to the halfway line of the other.

'Farmland' needs to be divided into land that is arable and land used for grazing animals. Arable land can be ploughed, and there is enough warmth and rain to grow crops. Grazing land is generally drier, colder or steeper, or land where the soils are thin and stony. On a global scale, one of our football fields is used for grazing only, and the other half a field grows crops, and planted pastures for animals.

Rice, wheat and maize are by far the most important crops for feeding the world. The next most important group are soybeans, potatoes, sorghum and cassava. If I planted the whole garden to potatoes and got a bumper crop each year, I still would not have enough to feed the family.

Energy cannot be created or destroyed. Every Calorie of food energy put in our mouths has to be converted to another form of energy. Most of the food energy keeps my body warm and functioning, or my maintenance requirement, and the rest is burnt during physical activity. If I eat more Calories than needed for maintenance and activity, the extra energy will be stored in my body as fat. If I eat less I will lose weight.

A lighter person has a lower maintenance requirement, but there are obviously limits to how little or how much we can eat. The average daily food energy intake in Mozambique is around 2000 Calories per day, and just over 3000 Calories in Australia. Countries consuming an average of 2000 Calories per person per day will, at least in some sectors of the community, experience persistent hunger, whereas countries consuming an average of 3000 have major problems with obesity.

The bulk of the world's food energy comes from just three plants, and they are all cereals. Leading the production statistics is maize (or corn) at 112 kg per person per year. Maize is followed by wheat (97 kg) and then closely by rice (94 kg). Coming in at a distant fourth is barley (24 kg), used mostly for making beer. Next is sorghum (9 kg), millet (4 kg), oats (4 kg) and rye (3 kg). Root crops also make a substantial contribution to our starch requirement, particularly potatoes and cassava at 10 and 9 kg per person per year respectively (or 51 and 31 kg if we do not adjust for the water these crops contain). Rice is actually the most commonly eaten food, because most of the world's maize and a substantial portion of the wheat are fed to animals.

A diet of cereals and root crops will probably not provide sufficient protein, apart from being downright boring. The main problem is that cereal protein usually does not have the full complement of the essential amino acids. Thus every culture supplements cereal protein as much as they can. This is either with legumes or animal products. By far the most commonly-grown legume is the soybean (32 kg each) but the vast majority of this goes into processed food and animal feed. The production of dry beans, chickpeas and lentils are all less than 3 kg per person per year. Fats and oils are essential as well, with rapeseed (canola), cottonseed, sunflower and groundnut being the main plant sources.

Animal sources are growing in importance, providing 17% of all Calories consumed by humans. Dairy products are the most important of these, followed by pork, poultry, beef and eggs. The global average meat consumption is 38 kg per person per year, but this varies enormously among countries. Indians derive only 1% of their food energy from meat, largely for cultural and religious reasons. Mozambicans also derive 1% of their energy from meat, but this is largely due to poverty. Their effort is directed to growing cereal and root crops, and there is no excess to feed to animals.

As countries get richer, people upgrade their diets. They want more meat. Australians get one-third of their Calories from animal sources, double the world average. In our household, we have made some minor attempts to reverse the trend, but it's hard to substitute vegetable protein for meat protein. Our attempts to slip broad beans into the diet has met strong resistance from the younger children, who have learnt to detect their presence even when they are ground up and disguised with various spices. I have eaten delicious and nutritious vegetarian meals in India, but we lack both the skill and time to do it ourselves on a regular basis.

We do produce animal protein in the form of eggs. Our five laying hens need about 300 Calories per day each, which adds up to about the same as our seven year old. We can expect four large eggs per day giving us a total of 340 Calories in return. The layer pellets we feed to hens are not a bad diet, being made up of various cereals, soybean and sunflower meal and sometimes meat meal as well. At 16% protein we could probably subsist on it ourselves, but eggs are much nicer. Herein lies both the problem and delight of living higher up the food chain. The hens require a maintenance ration just to keep themselves going, just like us. We put 1500 Calories per day into the hens but only get 340 Calories worth of eggs out the other side.

Since we only have from the goal line to the halfway line available for growing crops, we don't want to send too much of it through animals, especially when some people do not get enough cereal to eat themselves. On the other hand there are beneficial interactions between plants and animals. Animals can eat crop residues and make use of damaged or low quality grain. It is also good to rotate crop land and pasture land to retain fertility and control weeds and diseases. Half the feed requirements of our hens is produced from household wastes, including stale food and scraps from the table as well as weeds and crop residues from the garden.

There is also the other football field which is too dry, cold, steep or infertile for crops. Grassy plains and savannahs are composed largely of cellulose, the structural component of plants, which we cannot digest. Animals can, using the bacteria in their gut to do the initial breakdown of cellulose into sugar. Curiously the most commonly eaten meat is from pigs and poultry, rather than grass-eating cows, sheep and goats. This is a consequence of the higher food to meat conversion ratios in pigs and poultry housed under intensive conditions. At the low tech end of the farming spectrum, the omnivore status of pigs and poultry fits in better with small-scale operators on diversified farms.

The population is still growing, at least for the next few decades, and the squeeze is on our one-and-a-half football fields. If we factor in land degradation by soil erosion, soils going more acid or salty and nutrient exhaustion, and land lost to urbanisation, then the scenario looks a lot worse. Twenty-five years ago it would have been reasonable to say we were heading for disaster, but on gross averages alone, that's not what happened. Although around one in six of humanity lives out their days chronically short of food, the total energy intake has gone up by 10% in the last 25 years. In the most populous nations of China and India, the energy intake has gone up substantially more. So has the protein content of our diets.

When we look at the change in wheat yield in Australia over the last 100 years we see that yields have been increasing at around 1.5% per annum. That does not sound much, but it is compound growth, so it's 1.5% more of an increasing value. This modest annual increment takes average wheat yields from half a tonne per hectare to 2 tonnes per hectare, a fourfold increase in production in 100 years. The Green Revolution, which started half a century ago, was based around breeding cereal crops that were shorter and stockier. Some of the sugar that would have been turned into long stems and leaves ended up as starch in the grain. This period coincided with the widespread use of machinery, so the crops could be planted quickly just at the right time. The stronger stalks could hold up a larger seed head, so the crops did not become top-heavy and fall over when fertilised.

The Green Revolution has its critics. It is based around cheap fuel, particularly in the production of nitrogenous fertilisers. It can also be argued that the revolution did not deliver to the poorest of farmers, but when we look at nations as a whole, we see yield increases in countries like India and China equal to those of developed nations. The countries whose food situation has gone backwards are countries marred by internal conflict. Another criticism of the revolution is that it's over; we've got all the benefits and are now left with the associated problems. But this also does not appear to stand up to the available data. Yields are still increasing, though perhaps not as fast as four decades ago.

It looks like a lot of grain, but the world only stores enough food to feed us all for about three months. Food security means more food needs to be grown closer to where people live using less fossil fuel-based inputs.

The vast bulk of our plant food comes from just a handful of the 200 000 species out there. Of these, almost all are annual plants. Olive oil, palm oil and nut trees make a contribution in some diets, but overall it's a tiny fraction of what the annuals do for us. It is almost as though domestication and then the Green Revolution bypassed the perennials.

The reason for this is that the lifecycle of annual plants fits so well with agriculture. The annual plant matches its growth perfectly with the favourable growing season and survives the cold or dry time as seed. Large seeds, which store more energy and therefore provide the fastest getaway next season, are an advantage for the annual plant, but also for the farmer. When a wheat plant flowers and starts to make seed, it sucks most of the energy and protein from the leaves and stems and packs them away in the seed. The seed is a concentrated food source waiting for us to collect.

The strategy of the perennial is to survive the hard times, not pack their genes into a seed and sit them out. To survive, the perennials need deep roots to tap the last of the water and woody stems so that they can lift their canopies above their annual competitors. This takes energy that might have gone into seed production. And since perennials are generally not in the annual race to complete their lifecycle in one short rainy season, they do not go in for big seeds.

Perennials also have to survive greater pressures from parasites and grazers, because when the annuals are gone they have to face the pests alone. Many perennial plants produce thorns and chemicals to deter herbivores. This investment in infrastructure and defence diverts resources that could be used for harvestable new growth or seed. A perennial plant can usually capture more water, nutrients and light than an annual, but because of its deep roots and elevated crown, the harvestable proportion of digestible energy and protein is less.

In general, cereals do not to go in for chemical defences, which is one reason they are such an important food source. They germinate together when the wet season starts and therefore tend to flower and produce seeds en masse. No pest can get through such an abundance of food in a short space of time, so a large number of seeds are likely to survive predation. This 'go-for-broke' strategy means that cereals can put all their energy into making more seeds.

So far the role of the fruit and vegetable garden has disappeared from this discussion over global food requirements. Just 5% of the world's Calories come from fruit and vegetables. If we add in a few root and bean crops and the small livestock associated with gardens, the contribution would be much greater. What the garden does bring is variety and flavour, and the nutritional and health benefits of fruit and vegetables keep growing as we understand more.

Yet to keep feeding ourselves we will need to keep the yields of the staple crops increasing, and this will require more research and development than what brought us thus far. We also need to close the yield gap, the difference between what the top farmers achieve now and the average yields. This will require a massive effort in education and training. Since we currently eat only half the grain we grow (some is kept as seed for sowing with the rest fed to animals), diets will have to change.

There has been a trend to separate the farming of crops and the farming of animals, with more and more meat produced from intensively-housed animals eating grain and creating mountains of waste. There may be economies of scale here, but it makes little biological sense. As far as possible ruminant animals need to eat grass and crops need to be fertilised by leguminous pasture and manure. We will have to base more food production around perennial crops. The current way of doing things relies far too heavily on fossil fuel energy.

With only 16% of the world's cereals traded, more food will have to be grown where people live, particularly the poorest, who are not part of the international market. There are other formidable obstacles faced by farmers in poorer countries, from access to land to access to markets. It is within our grasp to reduce widespread and persistent hunger from planet earth, but by no means certain.

The recent worldwide rise in food prices gives us another wake-up call.

Chapter 17

To till or not to till

Annual plants provide most of our food, but we have to sow them year by year. We first need to get rid of whatever is growing in the field, and then produce a seedbed for the new crop. That means tillage. The plough has been used since ancient times; there are paintings of wooden ploughs dating back over 4000 years. The first plough was probably a forked branch. One fork was cut short and sharpened to gouge a furrow, while the other was attached to a draft animal.

The most familiar plough is the mouldboard. This plough had a curved cutting edge that sliced down into the soil and under the roots of whatever was growing. It was often drawn by a pair of oxen, and the farmer walked behind, holding two handles attached to the back of the plough. As the plough moved forward, the cut sod was lifted and turned over, so the plant growth was completely buried.

The mouldboard plough revolutionised farming in north-west Europe. Weeds or residue from the previous crop or pasture could be buried in one go, leaving fresh soil at the surface. It just took a raking or two to produce a seedbed. Manure could be easily incorporated and weed seeds were often buried to depths from which they could not germinate.

As agriculture moved to new continents, the mouldboard plough went along. The first problem with the mouldboard plough in Australia was that it kept hitting the tree roots and

The mouldboard plough turns the soil over, burying weeds and producing a seedbed. Animals pulled the first ploughs 4000 years ago and are pulling them still.

stumps of newly-cleared land. Many a plough was damaged until a clever man added a sharpened steel rod set just in front of the cutting edge of the plough and inclined forward in the direction of travel. When the bar hit a stump it rode over it, lifting the plough behind it, and then a counterweight brought the plough back to its correct position. It was called the stump jump plough, and it saved many farmers from ruin.

The real problem, however, was not losing the plough, but losing the soil. The mouldboard plough left the soil totally exposed to the wind and rain. This was not a problem for the rich organic soils under the gentle European rains, but it was a disaster in Australia and elsewhere. Erratic rains often came in heavy storms and the bare impermeable soils could not soak it up. It takes a thousand years for a centimetre of two of soil to form from the underlying rock, and just a few storms to wash it away.

New ploughs were invented that could kill weeds without disturbing plant material on the soil surface. This was particularly important for cropping land. Weeds proliferated when rain fell after harvest, and these had to be controlled or they would use most of the soil moisture for the next crop. Regular ploughing out of weeds used to leave the soil exposed to erosion for months between crops. The blade plough, for example, sliced just below the soil surface, severing weed roots but leaving the stubble on the surface to protect the soil. The seed was sown into a rough seedbed through the previous year's straw. As long as the seed was covered by moist soil, crops established and produced.

It soon became clear that of the three reasons for ploughing – killing weeds, making a seedbed and loosening the soil for root growth – killing weeds was the most essential. Many farmers converted to no-till farming en masse when herbicides became available so they could leave the residues to protect the soil surface. Cereals sown into untilled soils usually grew a bit more slowly than those planted into cultivated soil, but there were other advantages.

Apart from losing the protection of straw, those cultivating had one more operation to squeeze into the narrow window that represented the best time for sowing a crop. It could take many days to cultivate a farm after the opening season rains and the weeds started germinating well before the seed was planted. No-till farmers could plant as soon their tractors could get onto the land. The seeds went into harder soil, but the earlier sowing often compensated for the slower growth.

Cereal crops, like wheat and maize, have large seeds which make for a strong seedling that can put up with a rough seedbed. Their root systems can work their way a metre or more down into uncultivated soil. Vegetable crops are much more tender. Most have tiny seeds, so they need a fine seedbed to establish and soft friable soil to grow in.

Two of my favourite vegetable growing books that guided my early gardening activities had very different views on this. One proposed the double-dig method. First you dug out a trench of soil a spade deep at one end of the garden bed and then loosened the soil at the base of the trench with a fork. Then you dug a second trench, throwing the soil into the one you had just dug, and loosened the soil at its base. After digging the final trench, you had to fill it with soil from the first trench at the other end of the garden bed.

A double-dig garden was a lot of work, but the author had proved it was all worthwhile. The roots quickly permeated through the loose soil, easily scavenging for the water and nutrients needed for a bumper crop. Once you had a double-dig bed you never walked on it, so the benefits could last for several crops.

The other vegetable gardening book was adamant that you should never dig your garden. In fact digging soil was seen as destructive. You had to lay newspaper on the soil surface and then cover it with hay and compost. The newspaper smothered the weeds, the hay and compost decomposed to make a friable seedbed and the worms did the rest.

There is something nice about digging over a garden bed. Soft crumbly friable soil looks good and smells good. When you have to use all your weight to get the fork full distance into the soil, it feels like you are doing the right thing for the little roots that will have to force their way through such hard stuff in a few weeks' time.

There is of course a downside. Each time new soil is brought to the surface, it is exposed to the elements. A big raindrop hits a soil crumb with enough force to smash it apart. The crumbs of some soils collapse when the soil surface is saturated. The tiny clay particles get suspended in the water, leaving the larger silt and sand particles behind. When the soil dries the clay is deposited on the surface, forming a thin but hard crust that stops germinating seeds from coming up and water from going down.

The worst damage to soil gets done when it is ploughed too dry or too wet. This would not happen in a garden bed, because it is too difficult to dig very wet or dry soil, but it's not too difficult for big machines. The aim of ploughing is to break soil into crumbs, but a plough in wet soil smears and squashes the soil and a plough in dry soil shatters the soil into fragments.

Organic matter provides the glue that helps stick soil particles together. Without this glue soil would pack, a bit like 100 ping pong balls in a cardboard box. If we coated the ping pong balls with honey, and then moved the balls around, they would stick together giving a more 'open' structure. The honey acts to hold the balls apart, or rather to stop them from completely collapsing together. It is the same with soil.

Each time we bash the soil with a plough, we separate some soil particles that were glued together by organic matter. Micro-organisms live off the organic matter. When organic matter is stuck between two soil particles, the micro-organisms can't get at it, but when it becomes exposed by plough they have a field day. In the short term this is good. The decomposition of organic matter releases nutrients. In the longer term we lose the glues needed to hold the soil crumbs together.

Fragments of soil crumbs can block the existing pores in the soil. If we took 20 of the ping pong balls out of the cardboard box and ground them into small pieces, then returned them and shook the box, the ground up balls would fill the spaces between the existing balls. We would have the same weight of balls in 80% of the original volume. This is the conundrum of ploughing. At first we make the soil softer and less dense, but after many years of ploughing we predispose the soil to getting harder and denser than it was before.

This brings us back to the other method. We cover the soil with compost and straw, which protects the soil surface, provides ample organic matter and feeds the worms. Most people who have tried this method find it works very well. The problem comes when we want to scale up. We can plough many hectares of land in a day, but it would be hard to cover the same area with a thick mulch. Where would we get it from?

As a life-long vegetable gardener, this is the subject I wanted to tackle for my postgraduate studies. Fantastic work had already been done on growing cereals without tillage and many farmers had taken it up, but almost no commercial vegetable crops were grown without tillage.

Growing vegetables no-till seems logical enough, especially if you planted seedlings, and so did not need a fine seedbed. Since vegetable crops are usually grown in irrigated and well fertilised conditions, the roots would not have to scavenge far and wide like the rain-fed crops. The soil may be a little harder without cultivation, and the roots may grow a little slower, but does that matter when we are supplying them with luxury levels of water and nutrients?

Surprisingly, the answer depends on which crop we were growing. Tomato crops grown on no-tilled soil grew slower than their counterparts in freshly ploughed soil, but by the end of the season the fruit yield was the same. The no-till tomatoes were able to produce the same quantity of fruit from a smaller plant.

Lettuce did not perform as well under no-tillage. A lettuce crop, of course, is just leaves, so in that sense the result is the same as tomatoes. Storage crops, like beetroot, did quite well under no-tillage. They performed better than leafy lettuce and cabbage crops but not as well as the crops that fruit, like tomatoes and beans.

The no-till soils in the above experiments did not have any mulch to protect the soil surface or additions of organic matter to feed the worms and other soil organisms that burrow through soil. The next aim of the work was to find a way for no-till to work reliably for all vegetable crops. This meant cover crops.

Cover crops are crops grown especially to improve the soil. They may be legumes to add nitrogen, bulky crops to add organic matter, bio-fumigants that suppress disease or strong tap-rooted plants to punch holes into tough subsoils. Normally cover crops are ploughed under before the vegetable crop. The Clever Clover idea was a bit different, in that the cover crop died down on it own, so it could be followed by no-till planting. The clover system allowed us to test various aspects of what really improves the soil.

We started experiments on a piece of land that had not been cultivated for some time and was a bit compacted. On part of the area we grew the clover. Then, after the clover died down in early summer we removed the clover mulch from half the area and put it onto a no-till area that had not grown clover. We then ploughed some of the beds and put clover mulch, grown elsewhere, on half of these beds.

The experiment gave us six combinations. We had compacted soil and cultivated soil, each with and without a clover mulch (four treatments). Then we had compacted soil that had grown clover, but on half the area the surface mulch was removed and on half it was retained (two treatments).

Compacted soil with no mulch grew the worst crop, but the yield improved when the clover mulch grown elsewhere was applied to the surface. The mulch kept the soil surface a little moister, and moister soil offers less resistance to the growing roots. A lot of roots grow on the soil surface itself, just under the mulch.

Compacted soil also produced a better crop if the clover had been grown there the season before, even when the resultant mulch was removed. The roots of the clover had pushed their way through the soil, and, after decaying, left pores for the lettuce roots to follow. When we grew clover and left the clover mulch on the surface the growth was better still.

Cultivated soil produced better lettuce than the compacted no-till soil, but adding a clover mulch to cultivated soil gave variable results. In hot weather the mulch improved growth, but in cool weather it made it worse.

Growing vegetables between rows of lucerne proved to be a robust system. The lucerne produced all the fertiliser the crop needed and lots of mulch to protect the soil, making no-tillage a viable option. The lucerne picked up water and nutrients from the deep subsoil, extracting only a small amount from the adjacent vegetable rows.

Obviously there are a lot of things going on here. A mulch crop can improve the structure of a compacted soil, keep the soil a bit cooler and moister, but the early products of decomposing mulch are actually detrimental to the new crop. Everyone knows that fully rotted plant material is good for plants, but freshly killed plant material is full of chemicals that inhibit germination and the growth of other plants. These are called phytotoxic chemicals.

The length of the phytotoxic stage depends on how fast the material decomposes. Chopped up and buried material in moist soil rots faster than mulch on the surface. Some plants are much more phytotoxic than others, for example, decomposing ryegrass inhibited lettuce growth far more than decomposing clover.

The success of the Clever Clover idea therefore depended on whether the soil was compacted enough to need the clover roots to improve the soil, or whether the soil was getting too hot or dry. If the soil was already soft, cool and moist we were left only with the negative phytotoxic effects. This is why scientists should never give recipes to farmers.

Every scientist has a piece of work they wish they had done more of, and developed to its full potential. In my case it was the successor of Clever Clover, which was based around a rather different concept. I knew I needed mulch material, but growing it in rotation with vegetables proved problematic. Since I was not going to plough the clover under I needed to grow the perfect weed free cover crop. The next idea was to separate the mulch crop and the vegetable crop in space rather than in time.

Separation in space meant setting up long raised beds with furrows between. The rows were alternated; vegetable crops, then lucerne. Lucerne is one of those really remarkable plants. It does not grow more than half a metre tall, but its roots can go many metres down into the soil. Lucerne is a legume, grows fast and is very good at picking up water and nutrients from deep in the subsoil.

A most interesting thing about lucerne is that it does not send many roots sideways. Trees spread their roots deep and wide, but the lucerne plant goes for depth only. Perhaps the plant is so used to having a close neighbour that there is no point in looking sideways. It makes lucerne the perfect companion to grow alongside vegetables.

Vegetable crops were kept irrigated and fertilised and inevitably some water escaped past their shallow roots. The few sideways roots that the lucerne grew slowly picked up the water and nutrients leached below the vegetables. But the lucerne did not put enough roots sideways into the vegetable root zone itself to become a serious competitor.

We irrigated the vegetable rows with lines of drip tape and left the lucerne unwatered. It had to live off rain, and what the vegetable crop next door wasted. Every four to six weeks the lucerne was harvested and thrown across onto the adjacent vegetable bed. It also turned out that lucerne had no phytotoxic effects. It was the perfect mulch.

Perfecting the system means getting the right spacing between vegetables and lucerne and the right ratio of crop to mulch. We looked at various patterns; one row of vegetables and one of lucerne, or three and three or five and five. Wider spacing obviously reduced the competition, but it also reduced the capacity of the lucerne to reach the water draining below the vegetable rows.

The lucerne–vegetable system seemed to have everything going for it. We could produce enough mulch to go no-till and the lucerne hay provided all the nutrients and enough bulk to suppress the weeds. We protected and improved the soil and probably the ground water as well, but the system was hard to run on a commercial scale.

Modern tractors and spray booms cover multiple rows. It is impossible to spray one row and leave one, and the lucerne rows are simply in the way when it comes to harvesting machinery. This is the main reason why agriculture separates the pasture and crop phases in time, which we call rotation, rather than in space, such as the lucerne system.

After the formal experiments were over, we ran the lucerne–vegetable system very successfully in our home garden. Then, as the family grew and we needed to produce more food, we simply did not have the space for the lucerne. If we were ever to scale up again, and keep a milking goat or two to eat lucerne not needed for mulch, I would go back to this method of growing vegetables. Still, the conundrum remains. Production systems that are biologically more efficient than our current ones remain difficult to scale up.

Chapter 18

Permaculture to agroforestry

Every so often we are asked if a permaculture group can do a tour of our garden as the practical part of their permaculture course. I have never attended such a course, nor consciously planned our garden according to the principles of permaculture, but our garden apparently fits the genre. I am amazed at how many people enrol in such courses, and how keen they are to practise their craft.

Flip through a book on permaculture, and you see it is primarily about design, bursting with ideas on how to integrate our dwelling within a productive landscape. Like organic farming, permaculture is a reaction to conventional agriculture, entwined with a strong ethical and philosophical component. The core of permaculture is the reliance on perennial plants, arranged to take advantage of the intensity of use centred around the household and arranged to take advantage of every small niche that could provide a local advantage for this species or that.

Although annual crops are used in permaculture, the dominance of perennial species leads to a system of permanent agriculture. Hence the term 'perma-culture', as a contrast to our traditional agricultural systems based around the yearly routine of ploughing, sowing and harvesting. The word also carries the idea of stability over time. Permaculture sees agriculture as much more than the supply of food and fibre. It is a productive ecosystem encompassing as much diversity as possible to provide food, shelter and energy locally in a sustainable way, without a reliance on fossil fuels and agrochemicals.

If agriculture started by selecting and domesticating useful plants from the open woodland, and growing them in cleared areas close to the home, then permaculture seems to be heading back the other way. Permaculture aims to re-create the forest around our needs, but using domesticated, or partially domesticated species. That leaves a few questions for the agriculturalist. Are we to go back to being food gatherers again, albeit within a designed edible landscape? Can we feed the world this way?

Meanwhile the yield per hectare of annual crops increases year by year. Part of this yield increase is due to improved varieties. Part of the yield increase, at least in developed countries, is precision agriculture. Precision agriculture is about getting everything right for the crop; huge machines planting exactly in the right place at the right time, with the right amount of fertiliser. Satellite-guided tractors sow the crop in perfect alignment, using all available space. Harvesters record the yield in each part of giant paddocks as they reap the crops, so farmers can manage the variability across their farms.

Permaculture, with its focus on diversity and local niches, and precision agriculture, with a focus on large-scale uniformity, would seem to be the two ends of the spectrum. What lies in between?

In the early days of Australian agriculture, forests and woodlands were cut down for the plough or to allow more grass to grow for sheep and cattle. Tree clearing was encouraged by state and federal agencies and in some cases was a mandatory condition for occupying the land. This was not just a feature of pioneering agriculture. In the 1970s land was cleared at a rate of almost 1.5 million hectares per year. By the late '90s it had fallen to a quarter of that before recent laws put severe controls on further clearing.

Some farmers were moving against the trend. They made their money from annual crops or grazing stock, but had decided that they had lost too many trees. They believed they would be better off with more trees and so started revegetating the land. They were the first agro-foresters, 'agro' because they were essentially traditional farmers, and foresters because they were planting trees on their farms.

Essentially there were four reasons that sparked their interest in agroforestry. The first was that trees could bring benefits to their current practices. For example, trees produce shelter to sheep when lambing or after shearing. Some trees and shrubs can be used as animal fodder, and this can be the only green food about in the dry season or drought. Windbreaks protect crops from damage and excessive drying.

In all the above cases the trees had no financial value in themselves, they just added value to an existing enterprise; however, some farmers wanted to diversify their farms. Tree products like wood, pulp or oils could produce an income stream to buffer against the cyclical downturns in traditional farm enterprises like wheat and wool. This was the second reason behind agroforestry; to produce a new product.

The third reason for planting trees was to protect the longer term viability of the farm. Trees were planted to shield the soil from being blown away in the wind, or to prevent it from being washed away in the rain. The threat of dryland salinity generated a sense of urgency for revegetation on a massive scale. The felling of trees had changed the way water moved around the landscape, and salty water tables were rising closer to the surface. Trees were planted in an attempt to save the land in the long term.

The final reason behind planting trees on farms was for nature conservation. Trees were planted around remnants of natural vegetation to protect them from invasion by weeds. They were planted along creek lines and as corridors to connect up existing woodlands for wildlife. Trees added horizontal and vertical structure to the cleared plains, and in doing so provided new niches for other plants and animals.

The fascinating thing about agroforestry was that, whatever the primary reason for planting trees, there were other benefits to be gained. For example, we may decide to plant trees to counter dryland salinity (i.e. to protect the land). These trees could be arranged in such a way to connect up existing vegetation (nature conservation), to provide windbreaks (add value to existing enterprise) and even to be managed for their wood (new product). Agroforestry seemed like a new approach to farming, for a whole lot of good reasons.

There were examples from natural systems where trees and grass complemented each other, so why not agriculture? On the African savannah, for example, the grass seemed to grow better under or near trees. The local farmers tended to leave large trees in their fields cropped with maize, millet or sorghum. They did not have tractors, so the trees were no obstacle in that sense. But farmers reported that the trees actually benefited the growth of their crops.

The idea that trees and crops could be grown together was a revolutionary idea. In a dry country like Australia, chopping down too many trees had caused a problem. For most of the year crops and pastures could use every bit of rain that fell. But there were short times when more rain fell than the crops could use, like in the middle of winter when plants hardly used any water at all. This extra water slowly percolated down to the ground water.

Where the land was flat and the subsoils composed of clay, ground water makes very slow sideways progress to rivers. So the rain unused by crops started a slow rise of water tables over

The row of maize closest to the trees has died and the adjacent few rows have been strongly impacted by the trees. Any benefit from the tree component such as shelter or nutrients from pruning are outweighed by the competition.

huge areas. These water tables were once 20 m or more below the ground. Some started to rise at a couple of centimetres per year, others at tens of centimetres. As the water tables rose they collected salt from the subsoils, and carried it back up towards the soil surface.

Trees mostly prevented this from happening. Being long-lived, they could grow their roots a little deeper each year, sometimes reaching a depth of 20 m. Crops rarely managed to get roots past a metre and a half. Trees tended to use water from the topsoil in the rainy season, and if any water slipped past the shallow roots, it was picked up during the dry season by the deep roots. Very little water reached the water table, so they did not rise.

Cutting down trees for agriculture in places receiving lots of rain produced its own suite of problems. In tropical forests, there is far more rain than the trees can use, with the excess water giving rise to giant rivers like the Amazon and the Congo. Almost all the nutrients are held in the living trees and vines. When leaves fall and decompose on the forest floor, their nutrients are immediately sucked up by the surrounding trees before they can be leached to the rivers.

The slash and burn cultivators of the forest knew all about this. They felled and burnt a small area of forest, and then planted their crops. Crops were good for a few years as they lived off the nutrients released from the trees. But these did not last. The crops picked up some nutrients and the rains leached the rest away. Plots were abandoned and reclaimed by the forest, and a new area was cleared.

This shifting cultivation was thought to be more or less sustainable, as long as enough time elapsed between abandoning a cleared patch and felling it again. It took several decades for an abandoned patch to recover, and as the population increased, the clearing became too frequent. Eventually the forest failed to recover at all. This is where the idea started of growing trees and crops together.

Growing trees and crops together got the name of alley cropping. The trees or shrubs were grown in spaced rows, with the crops sown in the alleys between. Sometimes the alleys were only 5 m wide, between frequently pruned trees that looked like hedges. Sometimes the cropped alleys were 50 m wide between larger trees.

The first research trials on alley cropping came from wet tropical areas, and the results looked promising. Fast growing leguminous trees were planted with five rows of maize sown between them. The legume trees used water from the maize alleys between them, but in such wet climates this did not matter. What did make a difference was the pruning from the trees. The trees were cut back heavily at maize-planting time and the branches and leaves placed between the rows of maize. As they decomposed they provided slow release nutrients for the crop, so vital in these highly leached soils.

Alley cropping did not work as well in drier places like Australia. We were ready to discover all forms of complementarity between trees and crops. Trees, with their deep roots, could use water way below the root systems of the crop. They could also use water out of season, long after the annual crops were harvested. Trees could provide shelter, and recycle nutrients from deep in the soil to the surface by way of leaf fall. But though we looked for complementarity, we mostly found competition. When you plant a tree near a grass or cereal, and resources are scarce, the big guy usually wins.

As a general rule, it seemed that alley cropping only worked in a few specific locations. This left us with the conundrum as to why farmers in the African savannah left large trees in the cropped fields. Why did this seem to work?

It took a long time for the penny to drop. The large trees in the African fields had been there a long time. Their roots scavenged far and wide, so their competition for water in any one part of the field was low. Over the years many nutrients had been picked up from deep in the soil and returned to the surface in leaf fall. Much of the shade they threw was from branches and twigs, rather than leaves. Both branches and leaves throw shade, but it is the leaves that suck water from the soil.

It turned out we had chosen the wrong tree and shrub species for alley cropping. We wanted to have a lot of prunings, so chose species that grew very fast. In ecological terms, these trees tended to be the pioneer species that colonise new land after disturbance. Pioneer species grab all the water and nutrients they can, usually from the top soil where there is less work involved. We had inadvertently chosen highly competitive species, and then expected them to behave like mature old acacias on the savannah. It didn't work.

Some of the early enthusiasm for agroforestry has waned in recent years, but just as it seemed it was another failed hope, I heard a remarkable story. A friend of mine had been working in the Sahel, that narrow strip of land running along the southern boundary of the Sahara desert, cutting through countries like Mali, Niger and Chad. Farmers here eke out a

Informal gardens often have a mix of annuals and perennials, fruit and vegetables. These systems can be productive and flexible for local needs but are harder to scale up for full commercial production.

living, trying to match their short season millet and sorghum crops to the even shorter season rains. Even when the rains came, the crops were frequently lost to sand blasting and desiccation from the hot desert winds, or devoured by locusts.

My friend was planting trees by the tens of thousands. It was expensive business, and most succumbed to drought or goats within a year or two after being planted out. One day he looked out in some desperation over the endless fields that needed to be revegetated, but only saw weedy shrubs. On closer inspection the weedy shrubs were actually trees trying to re-establish after repeated slashing before each planting season. It was the sign of an underground forest. Maybe he did not have to plant trees, just protect what was already there.

Very slowly, a strategy of Farmer Assisted Natural Regeneration was born. As always it was pioneered by a few, who were ignored or ridiculed by the majority. But the few started to get results. The regenerating trees protected crops from the fierce desert winds, and the leaf litter made a huge difference to the fertility of the sandy soils. They could cut branches as fodder for their goats and as fire wood for sale.

There were fewer labour bottlenecks; the majority of work had been concentrated around sowing and harvesting the annual crop but the tree products could be harvested throughout the year. There was new habitat for birds and other predators and so smaller crop losses to pests. The women no longer had to spend hours searching for wood to cook the evening meal, so spent more time tending the crops.

Twenty years later a group of Dutch scientists were viewing satellite images of the area, and could not believe what they saw. The 1975 pictures showed a landscape with a few scattered trees but in 2003 there were trees everywhere. They estimated that 200 million more trees in the two to 25-year age bracket had appeared since 1975, spread over five million hectares of farm land. And they traced it back to the epicentre, the Farmer Assisted Natural Regeneration project.

There were many reasons for the outstanding success of this project. The project team had started with geometric patterns showing which trees should be planted where. But with Farmer Assisted Natural Regeneration you had to take what you got. All kinds of tree density, species composition and spacing were selected by farmers, depending on whether interests were towards cropping, firewood or fodder. Some farmers allowed the trees to get quite big in areas, then cut them for firewood and grew crops for several years as the trees resprouted before cutting for firewood again. The system was flexible to current and changing needs and responded to the varying prices farmers could get for wood, grain or meat.

The project also put enormous effort into farmer training, and showing farmers from different areas what others were doing. They ditched their recipes of what they thought would work in favour of on-going learning from whoever was willing to experiment. They even had to get the law changed so farmers own the rights to trees on their land. Most important of all they were in there for the long haul; worthwhile projects take about 20 years to really get going.

Some criticise this agroforestry project on the basis that it was geared towards a subsistence livelihood and neither commercial agriculture nor forestry. It would be hard to mechanise and needs a pool of cheap labour. Scaling up such systems is a real challenge. When we try to simplify agroforestry to an alley cropping system with straight rows of a single variety of tree with a crop in between, we seem to lose most of the complementarity between trees and crops, and are left with the competition.

The breakthrough for conventional agriculture will come when tree products have similar value to wheat and wool. This may come through multi-purpose trees providing fodder, wood, oils and biomass energy. The long time period between investment and return remains a major disincentive. Nevertheless, by the year 2000, 40 to 50 million trees were being replanted each year on Australian farms. About a third of this was commercial forestry, the rest largely for environmental reasons of one sort or another. The tide is slowly turning and the trees are coming back.

Chapter 19

The road to Tompi Seleka

Once a week we used to arrive at primary school with a small amount of money that was donated to a group called the Cape School Feeding Scheme. The teacher kept the tally at the bottom right-hand corner of the blackboard. At the end of the term the money was handed over to the Scheme and there was a competition to see which class had raised the most money.

Although we brought in some loose change every week, and knew the money went to provide a meal for school-going children, I never met any of these children. Throughout my entire school career I was not aware of meeting anyone who had experienced chronic hunger. Hungry people inhabited a different world and there was no connection between my world and theirs. I assumed the hunger of any school child in my city at the southernmost tip of Africa was assuaged by our contribution to the Cape School Feeding Scheme.

When I reached university I found a world of activists and idealists challenging the status quo. For me as a student of agriculture at Sydney University, my question became, 'why do so many people still go hungry in a world of plenty?' I read books and attended conferences and workshops on agricultural development. By the time I was half way through my degree I was so filled with naïve optimism that I just wanted to be let loose on the world.

My reading started to take me in another direction. What is 'development' and who gets to say who needs to be 'developed'? I recall the story of an aid group that was providing water to people in remote villages. They came across a man who voluntarily serviced village pumps in his area when he thought they needed it. The aid group saw this as such an important role that they paid the man to regularly service the pumps from project funds. When the project ended two years later the payments stopped and the volunteer moved on. With no one to look after them, the village pumps broke one by one, until the people had less water than they had before the aid project arrived. This simple story highlighted the biggest problem of all. You can be filled with good intentions, but can you actually do something good?

After leaving university I did some part-time volunteer work for an NGO focusing on third world development. My role, along with many other volunteers, was to review proposals for funding. In the early days, some of the proposals were hand-written and in very poor English. Nevertheless it was easy to see where the money should go. If someone had a sound idea, had connected with like-minded people and already made a start with the few resources they could scrape together, this was the beginnings of a project. This demonstration of courage and commitment was the essential ingredient. I discovered that development was about people long before it was about money.

Later on the proposals become more sophisticated. The benefit was that it became easier to see exactly how the money was going to be managed and spent, but the downside was that the recipients were learning all the buzzwords designed to hook the donors. It became harder to separate the courageous agents of change from those for whom aid was part of normal business. Nevertheless, after reviewing hundreds of projects I became convinced that an enormous amount of good was being done.

My own research career had started in Canberra when I came across these words:

> *We live in a world of exploding needs in which the most highly trained affluent individuals exchange their gifts in very small enclaves of wealth with others who can afford to pay for the products and services they provide.*

NGOs did fantastic work and it was good to volunteer some time and money, but I still operated almost exclusively within my own enclave of wealth. I wanted to break through to that other world, where farmers lived closer to the edge, but I had no idea how to do it.

I was working on a research project that I thought was relevant to far away resource-poor farmers, but I knew enough about agricultural development to proceed with caution. How do you know if the work is potentially relevant or useful in a place you have never worked and for a group of people you have never met?

Staring at the notice board at my place of work one day, I had an idea. A policy circular had just gone up saying that scientific sanctions against South Africa would be dropped, as the government was planning for the release of Nelson Mandela from prison. This was my chance to try and make a connection back to that 'other world'. The selling of Clever Clover kits had made a small profit and my sympathetic boss allowed me to put it towards a training scholarship. All I needed now was the right person to fill it.

The surprised agricultural attaché at the South African embassy listened to my request and then invited me out to lunch. I was trying to find a young professional who had grown up in a rural area, would have experienced hunger and deprivation, had studied agriculture at tertiary level and was now working to serve his people. A few weeks later the attaché said he had found my man.

When I picked up Maphala from Sydney airport, it was his first trip away from home. Maphala was very interested in the work. He was bright and a hard worker and we soon developed a mutual respect and friendship; however, I began to find out that this inhabitant of the 'other world' lived a life that was far more complicated than mine.

His family unit had broken down, partly through the migrant labour system. His parents were permitted to work in the cities far away but were not allowed to live there permanently as a family. Maphala stayed hundreds of miles away in a rural area with relatives where he had to stand up for his younger siblings to make sure they got a share of the food. His grandmother spent most of her meagre pension to help him through university. Now, in even further away Australia, he was trying to get his younger sister moved to a school that taught science and to get medical care for his ailing grandmother, both of which he had to pay for. The things I took for granted were mountains he had to move.

Maphala was a lecturer at the Tompi Seleka College of Agriculture, and after his return to the college from Australia I received a formal request to collaborate on some research projects there. Maphala and I scraped together a few small grants and got things started. It was the early 1990s and communication was still very difficult. There was no internet access and the mobile phone was still a rarity. The landline went through a manual exchange, and more often than not the copper wires had been stolen and melted down by small-time criminals. We communicated by snail mail and the occasional fax.

On my first trip to Tompi Seleka I was sure I was lost. The first 200 km north from Johannesburg was straightforward enough, but the last part of the journey seemed to be heading into uninhabited bush. Then suddenly the college, a 1500 hectare working farm with live-in accommodation for several hundred students plus classrooms and laboratories, rose out of the bushveld.

Tompi Seleka College of Agriculture was situated in one of South Africa's ex-homelands. It was an impoverished rural area from which many of the best and brightest had already left for the cities and the mines. The last town before reaching the College is the hub of a very productive fruit-growing district. Citrus and table grapes, as far as the eye can see on both sides of the road, formed the basis of a highly profitable export industry.

Then you drive through 40 km of largely uninhabited bushveld before reaching the College. If we continue heading north we travel on dirt roads through a dry and dusty place with villages of corrugated iron and mud shacks. From time to time you see a struggling maize crop, but it's mostly donkeys and goats picking their way through the thorny acacias.

My first question about Tompi Seleka was what do they teach? How to support the large-scale, high-tech, export-oriented agriculture 40 km to the south or how to improve the lives of the subsistence farmers to the north? Or is there some middle ground between these two extremes?

At the end of their studies, it was not clear whom a Tompi graduate had been trained to serve. Most were trained to be extension officers to the small-scale sector but there was precious little agriculture going on in the surrounding rural areas and the problems faced by the few farmers were usually beyond a young graduate.

The government used to employ most of the students as community workers of one sort or another, but even this was starting to wind down. During one of our visits a number of student grievances surfaced, from the quality of their food to their prospects of employment. A demonstration was planned. First the students went door to door, finding out who was for them and who was not. No middle ground was permitted. The students had always been friendly and mostly shy, but in demonstration mode the mood had shifted. Once the crowd had been assembled, the roads were blocked with burning tyres, and the march started with the curious toi toi-ing dance to the rhythmical protest songs. Students roughed up the vice-rector, ransacked some buildings, rounded up chickens, goats and pigs and let them loose in the main administration building.

This was my introduction to a world where the soaring aspirations of the people did not match the reality of their situation. Frustrations built up beneath the calm exterior, and these erupted into violence with seemingly minimal provocation. The demonstration died down as quickly as it had begun, and the next day the students behaved as if it had not taken place at all. Not so the pigs. They liked their new accommodation and had taken up residence in the tea room. It was only two days later that we managed to get them out.

After the riot, the pigs made themselves at home in the tea room. There was no lasting damage, but the lesson was clear. When people start to lose faith in the system of governance they can quickly take matters into their own hands.

My role at the college was to see if we could provide a way out of subsistence agriculture. When small-scale farmers grow a tonne of maize for their own household consumption, they save themselves the cost of purchasing it, with all the added costs of transport, milling, packaging and retailing. But the amount they grow beyond what they can eat in a year is worth a lot less. This must be sold, but the price for raw grain is set by global markets. They are effectively competing against the biggest farmers in the country and beyond.

This issue of scale is at the heart of the problem. In any country where agriculture is not subsidised, the farms are getting bigger. The only way farmers can survive is to spread their fixed costs, including that of their labour, over a much larger area. South Africa has a large rural population and not enough land or water for even a small minority to become large-scale farmers. And even fairly big-scale farmers are finding the going tough with low prices and a few tough seasons. So what hope is there for those struggling on a small scale?

The authorities' response to this dilemma was to form groups of small-scale farmers, provide them with land within an irrigation scheme, and so try and capture some of the economies of scale. For example the scheme had tractors to prepare the land, could purchase farm inputs in bulk and could market the crops. Ostensibly the small-scale farmer was in charge, and could purchase services from the scheme as needed. In reality, the scheme did more and more of the work until the only job left for the farmer was to move the irrigation pipes at the end of each day. Since the fields were some distance from where the people actually lived, it made sense for a few of the farmers to move the irrigation pipes for everyone. At the end of the season the farmers received a cheque for the crop sold and then paid most of it back to the scheme for services rendered.

These schemes were expensive to run, and essentially turned the small-scale farmers into part-time labourers. So at Tompi Seleka we decided to try and do the reverse. Instead of setting up schemes in places where there was already land and water, why not take water to

For every $1 spent on the military, just two and a half cents are spent on agricultural research. The prospect of famine on a huge scale probably poses the greatest threat to world peace, and in this case weapons will be of no use.

where people actually lived and try to grow much higher value crops. It may take more infrastructure to take water to people, but then we might see the water used in a much more creative way.

We set up a small production unit at the College based around the output of dirty water from a single low-pressure garden tap, which gravity fed from a dam. This ran night and day into a 4500 litre plastic holding tank, where some of the silt settled out. From there we used a tiny 400 Watt electric pump, feeding a system of drippers and micro sprays through a simple filter. Surprisingly electricity is available in some very remote villages.

Vegetable crops were grown using our no-till, mulch-based systems, so we did not need tractors or fertilisers. This was our attempt to use high-tech irrigation methods that did not have to operate at the large scale. The idea was to create a job for someone that paid better than some desk job in the town, with everything we needed to put the system together sourced locally and the produce sold locally.

The project made a great report for our donors. We had facts and figures showing it was all possible and pictures of our smiling students working hard. The problem was that when I was physically present at Tompi Seleka we made progress, but when I left, the progress stalled. No small-scale farmer took on our micro-irrigated, no-tillage, mulch garden production package, although some of the things we were doing did slip across the fence from the research plots to the commercial part of the college farm.

There were many other lessons to learn as an outsider living with people who had grown up in a different world. Since I spent a lot of my year stuck in an office at home, I loved getting out into the field and doing hands-on work. At Tompi Seleka I involved myself in all the physical work; planting, weeding, harvesting as well as taking all the plant and soil measurements. Sometimes I would need to go up to the central admin area, usually in my dirty field clothes.

One day I was taken aside and told that I was setting a bad example to the students. I may have done several degrees but as far as the students could see I had nothing to show for it except manual labour and old dirty clothes. It was then I realised that being a manual labourer was synonymous with being uneducated. Manual labourers were paid poorly and often poorly treated as well, and study was the ticket out of that world.

Now I understood the practice of my field assistant Moses. At the end of the working day Moses used to ask if he could borrow an iron. He then disappeared into the garage, took off his muddy clothes, ironed them, put them back on again, and sauntered back to the township.

I loved working at Tompi Seleka but it was here I learned that the solution to poverty is not, in the first place, a technological one. Poverty seemed to be a web of problems, and when one layer was removed another one took its place. If people had land they did not have water and if they had land and water they had no capital to start a business. And sometimes when there was surplus crop to sell, they could not get their produce to a market on time.

I see three ways to deliver agricultural aid. The first is to take ideas that have already worked somewhere and target new areas based on their similarities. There will always be unforeseen problems and the technology will have to be adapted, but the chances of some success are reasonable.

I was trying to follow the second way, trying to put together a complete new system of doing things in a difficult environment. To pull this off you need to commit a decade or so, and you need to be sure you are working on the right problem. The third way is to get involved in training and teaching.

I ended up getting involved in teaching. Tens of thousands of postgraduate students from developing countries have left their homes to study overseas. I wanted to do it the other way round. The postgraduate students would do their work closer to home – in Africa at least – and I would travel to work with them. I have not completed my own quest on the road to Tompi Seleka, the quest to find the ideas that will help the small-scale farmers turn less water into more food. But I have helped train a few African friends to join me on that journey.

Chapter 20

Simplicity

These days a scientist lives from project to project. We labour over lengthy funding proposals, promise breakthroughs, and itemise the steps that will get us there. We produce charts which show exactly how long each component of work will take, and what tangible output will prove we have completed each one. It is all about 'on time, on spec, on budget'.

It sounds fair enough. If we are receiving money to do a job, we need to be accountable. So we produce elaborate project plans together with monitoring, evaluation and risk strategies. At first it seems all this gives us some control over the business of scientific research, but this is not necessarily true. Sometimes there's a lot more to a problem than we first thought, and the rigid plans we come up with can make things worse in the long run. Science is not a machine where you feed resources and plans in one end and are guaranteed to get the right answer out the other.

My favourite example of unintended consequences comes from an attempt to ensure the survival of roan antelope in Kruger National Park. Roan, with their magnificent curved back horns and distinctive face markings, are a large antelope reaching over 250 kg. They used to be quite common, although they never reached the high densities of zebra and wildebeest.

Roan tend to favour areas growing taller, less palatable grasses. They wander over large areas of savannah, picking the few tender young leaves from the grassy tussocks. Zebras and wildebeest are more like lawn mowers. They concentrate in huge numbers on the more fertile soils which grow more palatable grasses, and by grazing patches of the savannah heavily, they ensure a continuous supply of nutritious young leaves.

Roan must cover large distances to select their diet, but they are also very dependent on water. There are large areas of the park with grazing suitable for roan, but without rivers for the roan to drink. So the park managers put in windmills to pump ground water into troughs. The plan worked spectacularly well. The roan spread out over a much larger area around the new sources of water. They still occurred at low densities because the grasses in the region tended to be of lower quality, but their overall numbers steadily increased.

Over the years, the new water attracted other grazers, particularly zebra and wildebeest. Although the dominant grass was not ideal for these species, they changed some areas with their constant heavy grazing, producing lawn-like areas of younger, more palatable grass. In doing so, the density of herbivores increased, and this did not escape the attention of the lions.

Since zebra and wildebeest are the favoured prey of lions, they get a lot of practice spotting predators and running to safety. Not so the roan. Hunting had been an unprofitable business for lions in the area prior to the new watering points, because of the low densities of prey. Now the lions switched their attention from the ever vigilant zebras to the unsuspecting roan. Roan numbers started to go down alarmingly.

The solution of adding new watering points was put into reverse. The bores were closed and, as predicted, the zebras drifted back to more favourable areas. As always the lions followed them, but not all the lions. A few prides took up residence in the roan areas. They changed their hunting habits. No longer could they hang around the plentiful herds and ambush them at waterholes on the short-grass plains. They started wandering widely, and met up frequently enough with the meandering roan to virtually wipe them out.

The story has two important lessons. First, what looks like the solution, and did work well in the short term, may not turn out to be the solution we had intended. It is true that more water was beneficial to the roan, but there were unintended consequences. Second, the response to the first 'solution' of adding water was not taking the water away again. In other words the road out of a problem may not be the same as the road that took us there.

The problem of the roan antelope is not an isolated one. In fact, in the area of natural resource management, this type of effect is seen more and more often. We call these complex problems, not so much because they are difficult to understand, but because the consequences of our actions are somewhat unpredictable.

It is very important to distinguish between problems which are difficult and those which are complex. If the problem is difficult, then we need to find clever people and give them enough time and resources to solve the problem. For example if our computer goes wrong, a skilled person can open up the innards and find out which particular circuit is at fault. There are thousands of components making up the circuit board, but each has a function and each behaves in a predictable way. By replacing the exact piece at fault, the computer will once again function as it should. It's a difficult problem, but solvable for an expert.

Complex problems may have bits that are difficult to understand, but the important thing is that they have an element of unpredictability about them. You can understand all the individual parts of the problem very well, but you can never be sure how the whole system will work when you put it back together again. The individual part, putting in watering points, gave roan antelope access to more grazing and increased their numbers, but at the system level even more roan got eaten.

Complex problems pose a real challenge to scientists in natural resource management, because it goes to the very heart of how we do our trade. Science is a systematic way of uncovering cause and effect to produce principles about the way the world works. We break problems down into component parts and study each in isolation to find out exactly what causes what.

For example: my tomatoes don't look as good as they should. The leaves are small, with some yellowing between the veins. One person thinks they have a virus. Another says I planted the wrong variety for this area. Yet another says it's because I planted tomatoes next to strawberries, or at the wrong phase of the moon. Some say it's because I'm not using organic principles, or the pH of the soil is wrong or it's because of the recent cold spell. Maybe it's tiny sucking insects. The effect – mottled leaves – is clear, but which of all the alternative causes above is the correct one?

Science is very good at dealing with problems like this. First we build on what is already known. For example we use a magnifying glass to look for mites on the underside of the leaf, or check the pH, and so narrow down the problem. After some preliminary investigation we may come to the hypothesis that the most likely cause, based on the visual symptoms, is a deficiency of the micro-nutrient zinc. We make up a solution of zinc sulphate and spray it on half the plants. If the mottling goes away on the treated plants but not the untreated, then we can draw a conclusion.

Real experiments are more formal than that, requiring replication and statistical analysis, to make sure we can separate some chance event from the real thing. But essentially science is about uncovering the real cause from several possible causes. Several of the possible causes can be interrelated. For example the tomato may have been deficient in zinc because the soil lacked zinc, or because the pH was so high that the zinc was insoluble. And if I had used an organic fertiliser like compost, it would have contained zinc so the problem would not have appeared in the first place. But the ultimate cause was zinc deficiency.

Science has been so successful that we know just about everything that can go wrong with a tomato. If a farmer gets low yields we can find out the reason and correct for it next time. But when it comes to a bigger problem, like how to farm sustainably in the long term, then there is a lot we don't know, largely because we are dealing again with complex problems.

Those who wrestle with complexity can find themselves at odds with the prevailing culture for science management. It is true that science has been spectacularly successful in solving problems, and in some fields new breakthroughs occur regularly. Because of this success, we scientists have become used to promising solutions to our funding masters in return for the cash we need to keep our laboratories afloat. We barely recognise when our promised solutions are no more than putting new water sources in the right places to save the roan antelope.

So how can we deal with complexity? One way is to simplify a problem as far as we dare to. When we simplify, we make explicit what we know and what we don't know. We aim for the simplest conceptualisation of the problem that all parties who want the problem solved can agree on; however, trying to simplify things can be dangerous territory for scientists. There is a well-known saying that, 'for every complex problem there is a simple answer that's wrong'. Mathematician and philosopher Alfred Whitehead said, 'seek simplicity and distrust it'.

Doctor turned author Spencer Johnson said that, 'it was during those years of looking at all of the complexities of medicine that I began to make a distinction between simplistic, which was not enough, and simple, which was everything it needed to be, but no more.' In a similar vein, the impressionist painter Hans Hofmann said, 'the ability to simplify means to eliminate the unnecessary so that the necessary may speak'.

It comes as no surprise that a mathematician, doctor and painter should have different views on the merits of simplicity, although we would think most scientists would side with Whitehead. Mathematics has right and wrong answers, but not all science is like that. When dealing with multifaceted problems, science is some combination of logic and art.

I got interested in the idea of simplifying problems when trying to advise irrigators about water. Precise measurements are needed to make sure a crop is well watered and I thought all irrigators should spend time and effort getting this right. After all, water is in short supply. But irrigators saw the problem differently. In order to run a profitable farm they had to get a lot of things right, not just water. There was crop nutrition, pest control, fixing machinery, managing labour and marketing – to name a few. The art of running a profitable business resides in getting the whole system roughly right, not one component exactly right. You could be the most accurate irrigator in the district and still go broke.

There was another issue I had not thought of. In my specialist area, the more detail the better; it may just help me to make a better decision. But in the context of running a whole farm, a lot of information on just one component does not necessarily help. For example I know that yield of my crop will increase with more water, but only up to a point. After a certain amount of water has been added, giving more does not increase yield. Giving far too much water decreases yield. But in between too little and too much there is a large space where, from the perspective of crop yield, it does not matter.

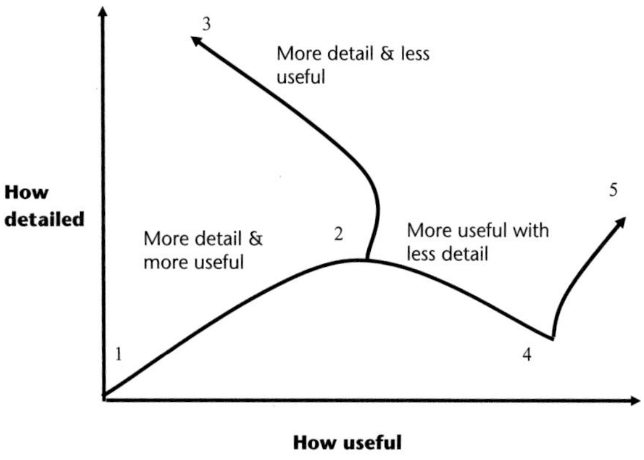

The simplicity cycle reminds us that adding more detail does not necessarily make something more useful. At some point we need to strip away the complicating detail and see the problem in a new light. The skill is to simplify without being simplistic.

I want the irrigator to be accurate, to apply not a drop more water after reaching maximum yield. But this adds risk to the farming operation. The fields are variable – so there is no one optimum point. And if we keep the farm just at the optimum point, and are hit by a heat wave or broken pipes, we can quickly slip into the danger zone. Moreover the cost of monitoring can outweigh the savings in water. An irrigator needs to find some workable balance between accuracy and managing risk and uncertainty. In other words they must balance the cost of extra information against the additional value it brings. What is the simplest way they can achieve this?

I came across a wonderful idea that has helped me to think through the relationship between more information or detail and its usefulness. We can think of something as simple as a bicycle. To make a useful bicycle there is some detail you have to get right, wheels, chains, brakes, etc. It is quite easy to get from position 1 to position 2 in the above figure and end up with a perfectly functional bike.

I rode to school every day on a bike that had three gears and never went to the repair shop. These days it's hard to get a bike with less than 12 gears, and without disk brakes, lightweight frames, suspension and many other added extras. For the long-range bicycle commuter or racer these added details add utility. But for me wanting to do the few kilometres to work each day, the detail has reduced the usefulness of my bike. There is invariably something broken, and I can't use any screw or bolt to fix it. I need the exact part from the specialist supplier and pay a fortune for it. My bike and I are firmly on the road to position 3.

This is a rather trivial example, and there are many like it, from video machines to automatic coffee makers. But it gets serious if we fall for the illusion that more detail always makes

I had to slow down to write this book, which started with placing myself in the garden and disconnecting from the internet for most of the day.

something more useful. Many working within large organisations have experienced this. Somehow the system demands more information, more forms, more meetings, more reporting but rarely provides anything more useful out the other end. Scientists working on complex problems can fall into the same trap. We get lost in the detail.

There are basically two ways to simplify. The first is to strip away some of the complicating detail and prepare the ground for new knowledge and insights. This essentially involves an about-turn on the journey to position 3, back towards position 2. Simplification comes when someone brings a fresh perspective that reduces the problem to its essential components and provides a more elegant way forward to position 4. From position 4 we may find that more detail is useful again, so we set off for position 5.

This stripping away of complicating detail does not work quite so well for the roan example. Because we don't understand all the factors that determine roan numbers, there is no obvious way of working out what detail we need and what is unnecessary. The second way to simplify is to find some integral behaviour of the system, something that tells us fairly easily how the whole system is responding to the major change of adding new watering points.

Just doing a roan census is not going to be enough, because by the time we are sure the numbers are falling, it may be too late to take action. Perhaps the best integral for understanding the impact of water points would have been to count the lions in the area as well as the roan. The number of lions tells us more about the whole system, because lion density changes in response to the total number of grazers. This would have given some early warning that the impact of the new water was going well beyond just the intended beneficiaries.

We are still left with the problem of how we go about simplifying. How do we find the fresh new perspective that takes us to the heart of the problem? It might be that we have to step back a bit from the planning and accountability structures that purport to make science more efficient. The quest for efficiency may be at the heart of the problem. We can think of this in terms of a journey.

If I'm sure of the destination and have produced the map of how to get there, I may as well travel quickly. I jump in a car, turn on the air conditioning, see the scenery rushing past, but focus on the destination. If I were on a bicycle, I feel the warmth of the air and hear the sounds of the bush on either side of the road. I would need to keep my eyes fixed on the road immediately ahead, but would be more aware of my surroundings.

When I walk, I take in the sounds and the scents, and I can also look much deeper into the bush on either side. It will take longer to reach a destination, but I may see the things the speedier traveller has missed.

The unnatural speed with which we live our lives compresses the world around us. We stare at the road directly ahead to avoid a collision, and so cannot absorb what surrounds us. Time gets compressed too. We do not sufficiently reflect on the past nor adequately contemplate the future. When travelling at speed we simply react to the immediate stimuli, in a mad rush to some destination we once convinced ourselves was the right one.

Slowing down is very difficult because it means we must acknowledge that we do not know, on our own, how to get to our destination. At best we have half the map, and the rest will have to come from others; those in our discipline, those in related disciplines and even those we may once have thought had little to offer. Slowing down means finding out what others already know, becoming a good learner and even unlearning some of the things that brought us to where we are now. Slowing down means allowing things to fail, because failure tells us what we most need to know.

Scientists must spend time immersed in the complicated detail, but we must also take time to pare away the unnecessary detail. The simplicity cycle encourages us to step back from the fast lane – after all, slowness may not be all that slow. Unreflective speed could well return us to where we began, or worse.

Chapter 21

Saving water

A visitor to our garden once confessed to feeling guilty about having a vegetable garden in the midst of the drought. Surely, in a water-stressed city, this was a luxury their family should do without. The logic is not quite correct. If you eat food then you consume water. Our city is no more short of water than anywhere else in the country. Of course we must ensure there will always be enough water in our city dams to drink, flush toilets and keep ourselves clean. But we have to eat too.

Anyone involved in food production, no matter how modest a scale, has some feel for the connection between water and food, and from time to time must wonder if they could do better. The truth is we can, if we look at it systematically, each step in the journey from water in the dam to food on the plate. This will show us where the inefficiencies lie, and what we could do about them.

Let's start with 100 litres of water in the dam, and use the most efficient distribution and irrigation system to water our crop. First, the water is piped from the dam to the farm. Major pipe infrastructures invariably have some leaks; a good system would get 95% of the water to the farm. The remaining 95 litres of water is then piped around the farm to each field, again at 95% efficiency. Five more litres are lost somewhere on the farm and the remaining 90 litres reaches the sprinklers in the field.

This foot pump can lift water from a shallow well and then to the crops. The pump could fill the tank in chapter 1 in about an hour. But it's fairly hard work, and a good incentive to use water wisely.

It is very difficult to get all the irrigation water into the crop root zone without some wastage, because soil, crops and irrigation systems are not uniform. The two major culprits are water running off the field without infiltrating and excess water draining below the root zone. We do well to get 85% of the 90 litres stored in the root zone, leaving 77 litres available for the crop to use.

Not all the water applied to the root zone is transpired by the plants. In fact, a lot evaporates from the soil surface before the plants can use it, particularly when the plants are small and the soil is not shaded. We would consider it excellent if 65% of the water was transpired and just 35% evaporated. The amount of water used by the plant is now 65% of 77 litres, or just 50 litres of the 100 that left the dam.

This is a scary exercise. It's only the water that goes through the leaves of a plant that helps us with food production. With a good piped infrastructure, a modern irrigation system and best practice by the farmer, only half the water is profitably used.

It gets a lot scarier when we look at a more typical situation. When the water is delivered to the farm through open, unlined channels, 20% of the water can easily be lost through evaporation and channel leakage before reaching the farm gate. On reaching the farm, the water may be stored in a small dam before being moved around the farm in earthen canals. Evaporation from the dam and leakage from the farm canals means only 70% of the water reaches the edge of the field.

Then we apply the water to the crop by flooding the field. It is difficult to get good uniformity with this method. Sometimes water runs off the far end of the field and it is common for more water to infiltrate into the top end of the field and drain below the roots. We may only get 60% of the water into the root zone, of which 50% evaporates from the soil surface. Only 17 of our original 100 litres are transpired by the plant. This example shows big savings are possible; moving from a poor system to a good system can give us three times the amount of useful water, from 17 to 50 litres.

There is a caveat here because not all the inefficiencies result in a net loss of water. For example the water leaking from canals can support useful vegetation on the banks. Water draining below the root zone of crops reaches the ground water and may find its way back to the river, or be pumped up and reused by another irrigator. One person's wastage can contribute to another's supply; however, water lost to evaporation, or reaching salty water tables, is completely lost to the irrigation water cycle.

So far we have only looked at the engineering aspects of getting water into the root zone. This is only half the battle. Now we have to get the biology right. Turning water into food is about exchanging carbon dioxide for water vapour in the leaves of plants, where the sugar produced eventually makes up the carbohydrate, protein and fats in our diet. To do this well requires a skilled farmer.

A well-grown crop can produce somewhere between one and two kilos of food from a thousand litres of water. Such productivity is only possible if we get everything right – no water or nutrient stress or attack by insects and disease. It's hard to do better than this because – so far – we can't get photosynthesis to work faster than it already does.

In contrast, a rather average crop may only produce 0.2 to 0.5 kg of food from 1000 litres of water. There are good reasons for this: infertile soils, inadequate water at certain times, poor farming techniques and pest attacks. Getting the basic things right can easily double, perhaps quadruple this productivity.

The majority of irrigation systems worldwide would lie in the poorer end of the spectrum. For these situations there is a potential to increase the water a plant actually uses threefold. And with the right kind of investment in farmer training and support we could increase the crop productivity of water threefold as well. Putting the engineering and the biological sides together gives us three times three or a ninefold improvement.

A 900% increase in productivity for irrigation seems an unbelievable number, but it is theoretically possible. In real life there will be all kinds of constraints that stop us getting there, but still there are points of intervention where big improvements can be made. We decided to test this idea at home, to search out every opportunity to squeeze as much productivity as we could from the water we use.

This has been a journey of gradual but constant improvement, around seven strategies. First the lawn had to go. I have some pangs of guilt over this, because my brothers and I destroyed my mother's garden through endless games of backyard cricket and rugby, and this necessitated a lawn that my own kids don't have. Giving up the back lawn is analogous to stopping some of the lower value production activities we get out of irrigation, something we will have to learn to do.

The next strategy was water tanks. There are opposing views on the merits of this. Many local governments give rebates to encourage households to install water tanks, but a straight economic assessment shows that tank water is not cheap. By the time you have purchased the tank and plumbed it into the gutters and then into the house or garden irrigation system, your water would cost three to five times more than that purchased from a desalinisation plant. Nevertheless I love my water tanks. A few wet days and we do not have to water the summer garden for a week, and then the tanks take us through a few more weeks of hot weather. I always leave some water in the tanks; I need water every day on demand for pots and seedlings, and the water restrictions do not allow me that. Then it's back to mains water for a while until the next shower of rain.

Strategy three is drip irrigation. Drip irrigation has two great advantages. It puts the water right next to the plant, so the wetted area of soil, from which evaporation takes place, is kept to a minimum. Second, drip irrigation applies the water in an extremely uniform way; the designs are so good these days that the last dripper on the line is scarcely different from the first. Uniform watering means an even crop, without the wet spots where water moves below the root zone and the dry spots that give us poor yields.

Strategy four is minimising evaporation from the soil surface. Drip irrigation is a very good start. As far as possible we use mulches, although they are not the panacea they are made out to be. I frequently see mulches laid so thick they would stop more rain reaching the soil than they would prevent in evaporation. One way we minimise evaporation is to concentrate light rainfall events from a large area to a small area. Rain falling on our glasshouse roof is caught in a gutter just above ground level. Perpendicular to the gutter run beds containing the deeper rooting crops like corn and pumpkins. The gutter has a corked hole at the head of each furrow and one is always left open. A light shower that would be useless for the garden is concentrated to just each furrow in turn, so that most of the water goes deep where it is protected from evaporation.

Strategy five is water harvesting during storms. All the crops are grown on beds with shallow furrows in between. Normally just a strip on the bed is kept wet through drip irrigation, with the edges of the beds and furrows dry. During heavy rain, water runs off the beds and collects in the furrows and the furrow becomes a storage zone for dry periods. When the furrow overflows the water is directed on the paths and low walls to a soakaway trench. The soakaway trench contains a large perforated pipe surrounded by sand with the base of the hole lined with plastic. It runs from the back garden, past all the grape vines and then past the fruit trees in the front. During wet periods the surplus water is diverted from the shallow-rooted vegetables to the deep-rooted vines and trees, where it can be drawn on for several weeks.

Strategy six is using grey water. Australian average household figures show that 20% of all domestic water is used in baths and showers and another 12% in the laundry. All our bath and shower water empties into a sump under the deck and a pump with a float switch immediately sends the water out to the garden. In very dry times I include the washing machine water. My concern is not so much the salt in the laundry detergent but the lint in the wash water that clogs my filters.

Strategy seven, the most important of all, is monitoring the amount of water the soil is holding. It takes some skill and experience to interpret the data, and almost none of the equipment available would be cost effective on a domestic scale or even for the small-scale farmer. Our garden is filled with my favourite gear, sometimes equipment sent to me for testing or cast offs from old experiments and of course various types of wetting front detectors.

Measuring the amount of water in the soil helps to ensure that the crop does not experience dry periods, so I can make sure I get the highest possible yields. Although a bit more difficult to interpret, monitoring can also tell me when I'm using too much water. There are other advantages too. Even when it's raining and the tanks are full, there are often parts of the garden where the soil below 50 cm is quite dry. I can store far more water in the plant root zones throughout the garden than in all my tanks combined. Frequently I irrigate during the rain to store water in my subsoils and so stop my tanks from overflowing.

I don't just monitor for day-to-day management. I monitor to learn. I keep a garden book to record which wetting front detectors are responding when, and the salinity and nitrate concentration of the water. I have a mental model of what should be happening, and when my data doesn't tally with my expectation, there is something to investigate and invariably something to learn.

When I first put in water tanks, I used watering cans to irrigate the beds. After just one can, the surface looks very wet, but I knew the water had only penetrated a centimetre or two. Each bed was watered in turn before returning to the first bed, giving some time for the water to soak in. After the second can there were a few small puddles in the low spots. After the third can little puddles joined up and water started to run into the furrows, so I never put on a fourth can. The soil surface looked wet, the plants were OK, but the wetting front detectors, which had popped up every second day under drip irrigation, never moved.

Tired from carrying water, I set up a holding tank two metres high, pumped water into it from the storage tanks and gravity fed the water into my drip lines. There were two immediate responses for the change from watering can to drip. First, the detectors 30 cm directly below the dripper started popping up again. Second, I only needed half the water, and the crop looked just as healthy. The experience just confirmed what I already knew, but sometimes you

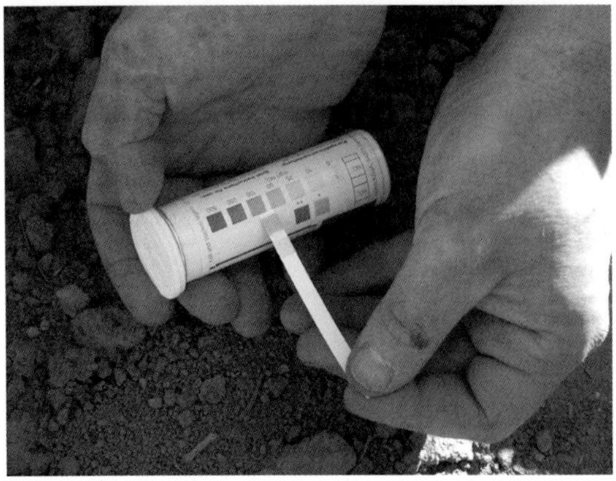

In the 40 years post-1960 when food production doubled, the use of nitrogen fertiliser went up eight times. Yet very few farmers measure nitrogen in the soil. Test strips are a quick and easy way to do it. When I saw how much nitrogen I was washing out of my soil I quickly changed the way I irrigated.

have to actually go through something to really believe it. You can save a lot of water by just putting the water where it's needed and restricting soil evaporation.

My journey of discovery took a leap forward when I started measuring what was in the water. By this I mean the conductivity of the soil solution, which is a measure of all the nutrients and salt dissolved in the soil water. I also started to measure the nitrates in the soil solution, because it's a good indicator of how quickly the plants will grow. Nitrogen, which is most commonly found in the soil as nitrate, is one of the five nutrients, along with potassium, phosphorus, calcium and sulphur, which all plants require in large amounts. Of these, nitrate is by far the most difficult to manage in soil because soil has no affinity for nitrate. Whereas other nutrients stick to clay particles or are sparingly soluble, nitrate just moves with the water. Over-water and the nitrates are quickly flushed out of the root zone.

Plant growth follows what we call a sigmoid curve which has the shape of a stretched out 'S'. At first they grow slowly, then they reach a point where they grow very fast until most of the sunlight is intercepted. Then they grow more slowly, because the lower leaves are shaded by the ones above. For example the plants might double in size every month. The small plants intercept 5% of the light in the first month, 10% after the second and 20% after the third. During the four and fifth month they intercept 40% then 80%, before slowing down as the light is fully used.

My gardening books told me I needed to prepare the soil well before planting, which included adding fertiliser or compost. Then the young plants had to be well watered to make sure the tender young growth was not compromised. I got a surprise when I started measuring the soil nutrients from the little water sample collected from the wetting front detector. The nitrate levels were through the roof.

Then later in the season, the wetting front detector stopped popping up. In week five I was giving double the water of week two, but that was not enough: the plants were four times the size. All too often, I discovered, I was applying too much water at the start of the season, and washing nitrate out of the soil, and too little water at the top of the 'S-shaped curve', the very time that yield is most susceptible to water stress. Let your tomatoes wilt when they are knee high and it hardly matters. Let them wilt at flowering and your yield can be cut in half.

Young plants do not need large amounts of nutrients, so loading up the soil before planting can be wasteful. Yet I was surprised to see high nitrate levels even without fertilising. Organic matter is always decomposing, and soil nutrients build up quickly when the soil is cultivated, warm and moist, and when there are no plants to absorb them. These are the very nutrients I was washing out with my obsession to be a good irrigator, at least an irrigator who likes to make sure the soil was always wet.

It is not realistic for most farmers to be constantly measuring water and nitrate levels, but from such experiences I have developed some simple rules of thumb. I put my detectors at two levels, around one-third and two-thirds the depth of the root zone. When the shallow ones trigger at the start of the season, I check the nitrate. If it's high, I reduce the water so that the shallow one, or more especially the deeper ones, are not triggered. If the nitrate is low I would add compost or manure as mulch, or apply soluble nutrients in the drip irrigation water.

As the growth rockets up the steep part of the 'S', I make sure the deeper detector responds from time to time. I'm no longer as concerned about leaching nutrients, most are already in the plant. But I am concerned about water stress, particularly around the flowering and early fruit growth stages.

The seven strategies we employ in our garden are, of course, local responses to local problems. I often pondered their wider value, and the usefulness of the anecdotes that go with them. Surely scientists should be able to diagnose problems and provide definitive solutions to farmers? But now I'm not so sure. I have been on a journey of learning how to grow more with less, just like every irrigation farmer with falling irrigation quotas.

These days when I walk onto a farm, the most important thing I can find out is what the farmer already knows. My job is to make explicit this experiential knowledge, derived from innumerable informal experiments and a lifetime of trial and error. Once we can articulate what we are trying to do and why, we can take the right measurements and structure our own learning.

Epilogue: Reflections in the scientist's garden

After being shortlisted for the fellowship to produce this book, I was asked to write a single page explanation of what it would be about. The final decision would rest on that.

The one-pager made several references to a health and diet book, written by eminent CSIRO scientists, which had become a run-away best seller. At the launch of my fellowship, the then Federal Minister for Science's speech picked up on this and pronounced that '*The Scientist's Garden* will do for water what the *Total Wellbeing Diet* book has done for food.'

But the Minister may have been disappointed because this book will not do for water what the *Total Wellbeing Diet* has done for food. For a start, this book has no recipes. This was perhaps my own lesson through writing this book. I had been searching for recipes where there weren't any. What I needed most was to become better at thinking through the issues and better at learning from others.

So the book has ended up about 'how to think about things', rather than 'how to do it' recipes. I hope the reader has been able to lay their own experiences over the stories in these essays and perhaps come to different conclusions.

The practical task of writing this book also had a lesson for me. I turned up to the office full of ideas ready to write, but thought it prudent to clear the email first. An hour or so later I felt drained by the hundred and one things demanding attention. So I got permission to write from home. I set up a desk where I could see the garden, and where I was not connected to the internet.

I viewed my mail offline to scan through the pages of messages, opened but still in the in-tray because at some stage I thought I needed to have another look. Buried but ignored among all the stuff that seemed so urgent on a day-to-day basis, were ideas, links and articles that someone or other knew would interest me.

The fellowship gave me the time to follow these leads and enjoy what others had done in a rather ad hoc way, and this gave me the stimulus to write. When I got tired of writing I worked in the garden. When I became disillusioned by the enormity of the problems and the smallness of my own contribution, I drew inspiration from the garden, which always brings forth its harvest in due season.

My own research career is laid out in these essays. I started with a completely different approach to irrigated horticulture, with a no-dig, legume-based rotation, interspersed with deep-rooted crops to pick up the leached water and nutrients. Then I joined up with many others who were looking to overhaul the way we do broad-scale dryland agriculture. The idea was to capture all the benefits of having trees in the landscape, with none of the downside for which they were removed in the first place.

Later I tried to make a soil water measuring device that was different from the ones that went before. It was designed to work in the same way people think about water – how deep did the water go into the soil? The idea was to make something that links water with the nutrients and the salt it contains, something intuitive that anyone could understand.

These days, scientists are under constant pressure to demonstrate the value of their research work, in order to justify further funding. I always find this difficult. Some days it seems we are on the cusp of a breakthrough, and other days the status quo seems immoveable. I see myself as sowing seeds (of ideas), and it is not always clear which ones will grow and how long they will take.

For this reason the life of a scientist must be characterised by a dogged perseverance combined with the continual search for new ways of looking at things. Perseverance is needed to stick with the ideas you believe in when the system around you is supportive and when it is not. New ways of looking at things are essential to stop you getting stuck in your own backwater.

A couple of years back I did a short working visit to Kruger National Park in South Africa. At the time I felt a bit guilty about it, because my topic of study there was not closely related to the work I was doing at the time. Yet the experience opened up an area of work that gave me unexpected new insights into the sustainability of food production, and linked me up with other scientists who have continued to influence my thinking.

The garden has played a similar role. It is partly an informal laboratory. I have filled many notebooks with measurements and observations from the various instruments scattered through the garden beds and between the vines and trees. These are not proper experiments, just a constant probing in an attempt to get the garden to give up some of its secrets. The garden is also the place to retire and reflect on what has been seen and heard. It's while doing the mundane jobs like weeding, that life slows down sufficiently to make sense of the day. It is the time to wrestle with different ways of looking at things. Ideas, like plants in the garden, need to be nurtured and given the space and time to grow.

Chapter notes

A book about feeding the world should have a very long reading list. Here I have just noted some of the sources that I used directly. Much of the material in this book was drawn from conversations with passionate people who love to share their ideas. I have listed some of my own work, much of which can be downloaded from www.fullstop.com.au.

1. The colour of water

Excellent statistics of water in Australia can be found on the ABS website at http://www.abs.gov.au/. See, for example, the *Water Account, Australia* (2004–05) (catalogue number 4610.0) and *Water use on Australian farms* (2006–2007) (catalogue number 4618.0).

For more information on water colours, see J Zygmunt (2007), *Hidden Waters: A Waterwise Briefing* (www.waterwise.org.uk), and UNESCO (2003), 'Water for People, Water for Life – UN World Water Development Report' (WWDR) (www.unesco.org/publishing).

2. A year in the vegetable garden

Up to the age of 15, I read every vegetable book I could lay my hands on, but I have to confess I have barely dipped into one since. A scientific text I use is L Pierce, *Vegetables: Characteristics, Production, and Marketing* (1987, John Wiley and Sons, New York).

3. Intrigue in the orchard

Two books I have greatly enjoyed reading on fruit production are L Glowinski, *The Complete Book of Fruit Growing in Australia* (1991, Lothian, Melbourne) and P Baxter and G Tankard, *The Complete Guide to Growing Fruit in Australia* (1981, Pan Macmillan, Australia).

4. Feathered thieves

The bird census data comes from the Canberra Ornithologists Group (2000) 'Birds of Canberra gardens' COG, and ACT Dept. Urban Services, Canberra and Greening Australia (2001) 'Bringing birds back'. I also used ideas from JT Du Toit, KH Rogers and HC Biggs (Eds), *The Kruger Experience. Ecology and Management of Savanna Heterogeneity* (2003, Island Press, London), and S McIntyre, J McIvor and K Heard, *Managing and Conserving Grassy Woodlands* (2002, CSIRO Publishing, Melbourne).

5. The lapsed organic gardener

I have shared countless cups of coffee with Joyce Wilkie and Michael Plane, organic market gardeners at Allsun Farm Gundaroo, NSW, wonderful friends and inspiring vegetable producers. For further information on organic production statistics see P Kristiansen, A Taji and J Reganold, *Organic Agriculture: A Global Perspective* (2006, CSIRO Publishing, Melbourne), and J Paull, 'China's organic revolution'. *Journal of Organic Systems*, 2007, pp. 1–11.

6. Outsmarting pests and disease

I took courses in entomology and plant pathology (insects and diseases) at university, but since then have had no formal study in these fields. I have only learned by necessity when under attack.

7. Flavour

An outstanding book, upon which much of this chapter is based, is B Patterson, *Fresh. Seeds from the Past and Food for Tomorrow* (1998, Allen and Unwin, St Leonards, NSW). A wonderfully accessible book full of interesting facts about the food we eat every day. Another is *Feeding People Is Easy* (2007, Pari Publishing, Pari, Italy) by Colin Tudge.

8. The tale of Clever Clover

I worked on the Clever Clover concept for about six years during the late 1980s, early 1990s. Some references on scientific aspects of the work can be found at:

RJ Stirzaker, BG Sutton and N Collis-George, 'Soil management for vegetable production 1. The growth of processing tomatoes following soil preparation by cultivation, zero-tillage and an in-situ grown mulch'. *Australian Journal of Agricultural Research*, vol. 44, 1993, pp. 817–829.

R Stirzaker, J Passioura, B Sutton and N Collis-George, 'Soil management for vegetable production 2. Possible causes for slow vegetative growth of lettuce associated with zero-tillage'. *Australian Journal of Agricultural Research*, vol. 44, 1993, pp. 830–844.

9. Soil and soil scientists

CSIRO has a wealth of brilliant soil and water scientists, and I have been privileged to work alongside some of them. For a summary of the tools irrigators use to measure soil water see R Stirzaker, 'Soil moisture monitoring and barrier to adoption'. *Irrigation Matters,* series No. 01/06, CRC for Irrigation Futures.

10. The Machingalana is talking to me

If you are interested in the dangerous journey of commercialising a piece of scientific equipment see R Stirzaker (2008), Should you commercialise your research? One scientist's story. https://rirdc.infoservices.com.au/items/08-169

More scientific treatment can be found at R Stirzaker, 'When to turn the water off: scheduling micro-irrigation with a wetting front detector'. *Irrigation Science*, vol. 22, 2003, pp. 177–185, and RJ Stirzaker, JB Stevens, JG Annandale and JM Steyn, 'Stages in the adoption of a wetting front detector'. *Irrigation and Drainage* (in press).

The website www.fullstop.com.au has all the information on wetting front detectors, including how to obtain them.

11. Watering grass

Thanks to the ACT local government for letting me dig holes and string wires under some of their ovals. More details can be found at R Stirzaker 'Saving water on sports ovals in Canberra'. *Report to Canberra Urban Parks and Places*, 2004.

12. Grey water and the washing machine

What you really need to know is which product to use, and that can be found at www.lanfaxlabs.com.au, with a lot of other good information. See also *Choice Magazine* (March 2007) for comparisons on how well different products do the job: www.choice.com.au.

13. At the end of the river

More information on the Murray–Darling Basin can be found at http://www.mdbc.gov.au/ and http://www.environment.gov.au/water/mdb/index.html.

14. The Goldilocks principle

I am impressed and thankful for the work of the Angas Bremer grape growers, and particularly the role played by Tony Thomson. Much of the information these growers collected can be accessed at www.angasbremerwater.org.au/.

The idea of the Goldilocks variable was adapted from T Lynam and M Stafford Smith, 'Monitoring in a complex world – seeking slow variables, a scaled focus and speedier learning'. *African Journal of Range & Forage Science,* vol. 21, 2004, pp. 69–78.

15. A short history of agriculture

The beginning of this story is based around J Diamond, *Guns, Germs and Steel: A Short History of Everybody for the Last 13000 Years* (1997, Vintage, London), and I have borrowed ideas from colleagues John Passioura, Tony Fischer and John Angus. For an excellent summary on productivity growth in Australia see J Passioura, 'Environmental biology and crop improvement', *Functional Plant Biology,* vol. 29, 2002, pp. 537–546. Costs of land degradation come from SA Hajkowicz and MD Young (Eds), 'Values of returns to land and water and costs of degradation: a consultancy report to the National Land and Water Audit', (2002, CSIRO Land and Water, Canberra). See also B Walker and D Salt, *Resilience Thinking: Sustaining Ecosystems and People in a Changing World* (2006, Island Press, Washington DC) for a critique on the search for more efficiency.

16. What we eat

Global food production and consumption statistics can be found on the FAO website at: http://faostat.fao.org/default.aspx.

17. Tillage and no-tillage

There is a vast literature on no-tillage. More information on my small contribution of no-tillage for vegetables production can be found in R Stirzaker and I White, 'Cover crop amelioration for no-tillage vegetable production'. *Australian Journal of Agricultural Research,* vol. 46, 1995, pp. 553–568 and R Stirzaker, 'The problem of irrigated horticulture: matching the bio-physical efficiency with the economic efficiency'. *Agroforestry Systems,* vol. 45, 1999, pp. 187–202.

18. Permaculture to agroforestry

Again there is a vast literature on agroforestry and a growing literature on permaculture. Some of my agroforestry work is contained in N Abel *et al.*, *Design Principles for Farm Forestry* (1997, RIRDC, Canberra) and RJ Stirzaker, R Vertessy and A Sarre *Trees, Water and Salt: An Australian Guide to Using Trees for Healthy Catchments and Productive Farms* (2002, RIRDC, Canberra).

African examples are drawn from C Ong and R Leaky, 'Why tree-crop interactions in agroforestry appear to be at odds with tree-grass interactions in African Savannahs'. *Agroforestry Systems,* vol. 45, 1999, pp. 109–129; and T Rinaudo, 'The development of Farmer Managed Natural Regeneration'. *LEISA Magazine,* vol. 23, 2007, p. 2.

19. The road to Tompi Seleka

Quote from Tom Sine's book, *The Mustard Seed Conspiracy* (1981, W Publishing Group, Nashville, USA). Thanks to the Tompi Seleka College of Agriculture for their collaboration over a period of several years. Since then I have carried on the work with John Annandale, Joe Stevens and Martin Steyn from the University of Pretoria, together with many students.

20. Simplicity

My time at Kruger National Park expanded my scientific horizons and I am incredibly grateful to Harry Biggs and Dirk Roux for being so generous in sharing ideas and insights. The ideas on simplicity are based on D Ward (2005), *The Simplicity Cycle. Simplicity and Complexity in Design* (www.dau.mil/pubs/dam/11_12_2005/war_nd05.pdf), and of slowing down on P Cilliers, 'On the importance of a certain slowness'. *Emergence: Complexity & Organization*, vol. 8, no. 3, 2006, pp. 106–113.

21. Saving water

The water use efficiency figures are based on T Hsiao, P Steduto and E Fereres, 'A systematic and quantitative approach to improve water use efficiency in agriculture'. *Irrigation Science*, vol. 25, 2007, pp. 209–231.

Many of the pictures in the book come from our own household collection. A good number come from friends who happened to be on the spot at the right time (thanks to Joyce Wilkie, David Salt, Ted Lefroy, Tony Thomson, Bertus Kruger, John Annandale and Felix Reinders) and from CSIRO colleagues (including Heinz Buettikofer, Greg Heath, Willem van Aken, John Manger, Tony Nadelko and John Coppi). Thanks also to Chris Taylor, Ash Leber, Robin Snow and Brad Collis who filled in a few gaps.